IEEE 802.11ba

IEEE 802.11ba

Ultra-Low Power Wake-up Radio Standard

Steve Shellhammer
Alfred Asterjadhi
Yanjun Sun

Published by
Standards Information Network

IEEE PRESS

WILEY

Library of Congress Cataloging-in-Publication Data
Names: Shellhammer, Stephen J., author. | Asterjadhi, Alfred, author. |
    Sun, Yanjun (Engineer), author.
Title: IEEE 802.11ba : ultra-low power wake-up radio standard / Stephen Jay
    Shellhammer, Alfred Asterjadhi, Yanjun Sun.
Description: Hoboken, New Jersey : Wiley, [2023] | Includes bibliographical
    references and index.
Identifiers: LCCN 2022049546 (print) | LCCN 2022049547 (ebook) | ISBN
    9781119670957 (paperback) | ISBN 9781119670995 (adobe pdf) | ISBN
    9781119670902 (epub)
Subjects: LCSH: IEEE 802.11 (Standard) | Radio–Receivers and reception.
Classification: LCC TK5105.5668 .S54 2023 (print) | LCC TK5105.5668
    (ebook) | DDC 004.67–dc23/eng/20221107
LC record available at https://lccn.loc.gov/2022049546
LC ebook record available at https://lccn.loc.gov/2022049547

Cover Design: Wiley
Cover Image: © Jackie Niam/Shutterstock

Set in 9.5/12.5pt STIXTwoText by Straive, Pondicherry, India

Contents

Author Biography *xi*

1   **Introduction** *1*
1.1   Background *1*
1.2   Overview *3*
1.3   Book Outline *5*

2   **Overview of IEEE 802.11** *9*
2.1   Introduction *9*
2.2   Overview of the IEEE 802.11 PHY Layer *10*
2.2.1 Operating Frequencies and Bandwidths *10*
2.2.2 OFDM *11*
2.2.3 OFDM PPDU *12*
2.3   Overview of IEEE 802.11 MAC Layer *16*
2.3.1 Network Discovery *16*
2.3.2 Connection Setup *18*
2.3.3 Coordinated Wireless Medium Access *19*
2.3.4 Enhanced Distributed Channel Access *20*
2.3.5 Security *20*
2.3.6 Time Synchronization *21*
2.3.7 Power-Saving Mechanisms *21*
2.3.8 Orthogonal Frequency Division Multiple Access
       (OFDMA) *23*
2.4   Conclusions *24*
       References *24*

3   **Wake-up Radio Concept**   *25*

3.1   Introduction   *25*

3.2   Primary Sources of Power Consumption in an IEEE 802.11 Station   *26*

3.2.1   Power Consumption in Transmit Mode   *26*

3.2.2   Power Consumption in Receive Mode   *28*

3.2.3   Power Consumption in Sleep Mode   *30*

3.2.4   Power Consumption in Deep Sleep Mode   *30*

3.3   Wake-up Radio Concept   *31*

3.4   Example of Power Consumption Using a Wake-up Radio   *37*

3.5   Selection of Duty Cycle Values   *39*

3.6   Conclusions   *42*

4   **Physical Layer Description**   *43*

4.1   Introduction   *43*

4.2   Requirements   *45*

4.3   Regulations   *47*

4.4   Link Budget Considerations   *50*

4.5   Modulation   *53*

4.6   Physical Layer Protocol Data Unit (PPDU) Structure   *55*

4.6.1   Non-WUR Portion of PPDU   *55*

4.6.2   Sync Field   *58*

4.6.3   Data Field   *61*

4.7   Symbol Randomization   *62*

4.8   FDMA Operation   *66*

4.8.1   40 MHz FDMA   *66*

4.8.2   80 MHz FDMA   *67*

4.9   Additional Topics   *67*

4.10   Conclusions   *68*

      References   *68*

5   **Physical Layer Performance**   *73*

5.1   Introduction   *73*

5.2   Generic Non-coherent Receiver   *73*

5.3   Simulation Description   *75*

5.3.1   Transmitter Model   *76*

5.3.2   MC-OOK Symbol Waveform Generation   *76*

5.3.3   Channel Model   *77*

5.3.4   Receiver Model   *79*

5.3.5   Performance Metrics   *80*

5.4   PHY Performance: Simulation Results   *81*

5.4.1   Sync Field Detection Rate   *82*

5.4.2   Sync Field Classification Error Rate   *83*

5.4.3   Sync Field Timing Error   *85*

5.4.4   Packet Error Rate   *88*

5.4.5   Effects of Transmit Diversity   *88*

5.5   Link Budget Comparison   *92*

5.5.1   Comparison to the 6 Mb/s OFDM PHY   *93*

5.5.2   Comparison to the 1 Mb/s Non-OFDM PHY   *94*

5.6   Conclusions   *95*

References   *95*

6   Wake-up Radio Medium Access Control   *97*

6.1   Introduction   *97*

6.2   Network Discovery   *97*

6.2.1   General   *97*

6.2.2   WUR Discovery   *98*

6.3   Connectivity and Synchronization   *102*

6.3.1   General   *102*

6.3.2   WUR Beacon Frame Generation   *102*

6.3.3   WUR Beacon Frame Processing   *104*

6.4   Power Management   *105*

6.4.1   General   *105*

6.4.1.1   MR Power Management   *105*

6.4.1.2   WUR Power Management   *106*

6.4.2   WUR Modes   *108*

6.4.2.1   WUR Mode Setup   *108*

6.4.2.2   WUR Mode Update   *110*

6.4.2.3   WUR Mode Suspend and Resume   *111*

6.4.2.4   WUR Mode Teardown   *111*

6.4.3   Duty Cycle Operation   *112*

6.4.3.1   WUR Duty Cycle Period   *113*

6.4.3.2   WUR Duty Cycle Service Period   *114*

6.4.3.3   WUR Duty Cycle Start Time   *114*

6.4.4   WUR Wake Up Operation   *116*
6.4.4.1   Individual DL BU Delivery Context   *116*
6.4.4.2   Group Addressed DL BU Delivery Context   *119*
6.4.4.3   Critical BSS Update Delivery Context   *121*
6.4.5   Use of WUR Short Wake-up Frames   *124*
6.4.6   Keep Alive Frames   *126*
6.5   Frequency Division Multiple Access   *127*
6.6   Protected Wake-up Frames   *129*
6.7   Conclusion   *130*

7   Medium Access Control Frame Design   *131*
7.1   Introduction   *131*
7.2   Information Elements   *131*
7.2.1   General   *131*
7.2.2   Elements Supporting MR Functionalities   *132*
7.2.2.1   DSSS Parameter Set Element   *133*
7.2.2.2   EDCA Parameter Set Element   *133*
7.2.2.3   Channel Switch Announcement Element   *135*
7.2.2.4   Extended Channel Switch Announcement Element   *136*
7.2.2.5   HT Operation Element   *136*
7.2.2.6   VHT Operation Element   *137*
7.2.2.7   Wide Bandwidth Channel Switch Element   *138*
7.2.2.8   Channel Switch Wrapper Element   *139*
7.2.2.9   HE Operation Element   *139*
7.2.3   Elements Supporting WUR Functionalities   *142*
7.2.3.1   WUR Capabilities Element   *142*
7.2.3.2   WUR Operation Element   *142*
7.2.3.3   WUR Mode Element   *145*
7.2.3.4   WUR Discovery Element   *154*
7.2.3.5   WUR PN Update Element   *155*
7.3   Main Radio MAC Frames   *155*
7.3.1   Beacon Frame   *155*
7.3.2   Probe Request/Response Frames   *156*
7.3.3   (Re)Association Request/Response Frames   *156*
7.3.4   Action Frames   *157*
7.4   WUR MAC Frames   *157*
7.4.1   WUR Beacon Frame   *161*
7.4.2   WUR Wake-up Frame   *161*

7.4.3   WUR Discovery Frame   *164*
7.4.4   WUR Vendor-Specific Frame   *165*
7.4.5   WUR Short Wake-up Frame   *166*
7.5   Conclusion   *167*

Index   *169*

Author Biography

Steve Shellhammer is a Principal Engineer and Manager in the Qualcomm's Wireless Research and Development division, working on new IEEE 802.11 standards. These standards include IEEE 802.11be (Extreme High Throughput), IEEE 802.11az (Enhanced Positioning), and IEEE 802.11bf (RF Sensing). He was the PHY ad hoc chair for the IEEE 802.11ba Wake-up Radio Task Group. He initiated and led a project to develop a prototype ultra-low power wireless wake-up receiver. He recently has been leading a project on RF Sensing at Qualcomm, focusing on RF Sensing applications. In the past at Qualcomm he led a cognitive research project for wireless networks.

He is currently the chair of the IEEE 802.19 working group on wireless coexistence and a member of the IEEE 802 Executive Committee, overseeing development of new IEEE 802 standards. He was previously a member of the IEEE-SA Standards Board and the IEEE-SA Review Committee. He was also the chair of the IEEE 802.15.2 Task Group on wireless coexistence addressing coexistence between Wi-Fi and Bluetooth. He was also the Spectral Sensing lead for the 802.22 wireless regional area networking working group.

Before joining Qualcomm, he worked as a Wireless Architect at Intel's wireless local area network division. Prior to that, he was the director of the Advanced Development department at Symbol Technologies. He was also an adjunct professor at SUNY Stony Brook, where he taught graduate courses in electrical engineering. These courses focused on probability, linear systems, digital signal processing, communication theory, and detection theory.

He has a PhD in electrical engineering from University of California Santa Barbara, an MSEE from San Jose State University, and a BS in physics from University of California San Diego. He is a senior member of the IEEE.

Alfred Asterjadhi is a Systems Engineer in the Qualcomm's Wireless Research and Development division. His focus is on the design, standardization, and performance evaluation of MAC protocols in wireless systems. Special emphasis has been given to increasing efficiency, scalability, and reducing power consumption in wireless networks. He is the chair of IEEE 802.11be, and has served as the vice chair of IEEE 802.11ax and the vice chair and technical editor of IEEE 802.11ah. He has made significant technical contributions to these amendments in both standardization and certification of Wi-Fi interoperability. He has authored and coauthored multiple conference and journal papers in wireless communications, a book in underwater acoustic networking, and is the holder of several patents in wireless communications. Alfred Asterjadhi received his PhD degree in information engineering and BSc and MSc degrees in telecommunications engineering from the University of Padova, Italy.

Yanjun Sun is a Senior Staff Engineer in the Qualcomm's Wireless Research and Development division, working on new IEEE 802.11 standards.

He received his PhD degree in computer science from Rice University and was a research engineer in the DSP systems R&D center at Texas Instruments before joining Qualcomm.

His research interests include systems design and standards development related to Wi-Fi, Bluetooth, Zigbee, protocol stacks, operating systems, and embedded systems. He has published papers at top IEEE and ACM conferences with best paper awards and driven R&D ideas into best-selling products in the market.

1

Introduction

1.1 Background

The Institute of Electrical and Electronic Engineers (IEEE) is a major professional organization which, among other things, develops standards. One of the most famous IEEE standard is IEEE 802.11, which is a standard for wireless local area networks. Implementations of IEEE 802.11 number in the billions. IEEE 802.11 implementations are certified by the Wi-Fi Alliance and referred to as Wi-Fi products in the marketplace. These Wi-Fi products have become ubiquitous and are used for wireless connectivity for smartphones, laptop computers, tablet computers, and numerous other devices. Wi-Fi networks are used in homes, business enterprises, and many other locations.

Since the first version of the IEEE 802.11 standard published in 1997, there have been many amendments to the standard which provide additional capabilities. An amendment to the standard adds content to the standard specifying new features and capabilities. Some of these are well-known major amendments like 802.11n, 802.11ac, and 802.11ax, which bring significant increases in data rate and network throughput. There are also smaller amendments to the IEEE 802.11 standard, which are more focused on a few new key features.

Many of these new amendments to the 802.11 standard, which increase the data rate and throughput of the network, often also increase the power consumption of the 802.11 devices. This is because either additional electronic circuits are needed to implement these new higher-rate features or the electronic circuits need to operate at a higher clock rate, both of which can lead to an increase in power consumption. This is the trade-off that comes with these increases in network speed and throughput.

IEEE 802.11ba: Ultra-Low Power Wake-up Radio Standard, First Edition.
Steve Shellhammer, Alfred Asterjadhi, and Yanjun Sun.
© 2023 The Institute of Electrical and Electronics Engineers, Inc.
Published 2023 by John Wiley & Sons, Inc.

There are a number of methods to reduce power consumption in 802.11 devices. Most of them focus on allowing the device to enter a power save (PS) mode where some of the electronic circuits are not needed at the moment and hence lead to the power saving. However, often the 802.11 device needs to exit PS mode periodically to allow it to transmit and/or receive data packets. Therefore, there are limits to how much power savings can be obtained using these power savings techniques. So it became apparent that there was a need for a new amendment to the 802.11 standard to focus specifically on power savings and to use a more aggressive power savings technique to enable more significant power savings in 802.11 devices. This led to the development of the IEEE 802.11ba amendment of the 802.11 standard.

The IEEE 802.11ba, which was published in 2021, is a standard for a wake-up radio (WUR). A WUR is a low-power wireless technology that enables the development of ultra-low-power devices. The term "radio" is used here as a synonym for "wireless." A WUR is a wireless technology that supports implementation of an ultra-low-power wake-up receiver, which has limited wireless functionality but can be operated at very low-power consumption levels. An 802.11 device that includes this wake-up receiver can allow its main transmitter and receiver to be placed into a deep sleep mode, while the wake-up receiver can be used to receive messages from an 802.11 access point (AP) when information is to be sent to or received from this 802.11 device. When the WUR receives a wake-up message, then the main wireless technology, or main radio (MR), in the devices can be brought out of its deep sleep mode and exchange messages with the AP. This MR can be any of the standard 802.11 PHY layers, including 802.11n, 802.11ac, 802.11ax, and so forth.

One can think of this as a hierarchy of wireless systems where the MR is used to transfer data, often at a high data rate (HDR), and the WUR is used to maintain a connection between nodes in the wireless network and wake up the MR when it is time to transfer data. This hierarchy of wireless systems provides both low-power consumption and the ability to transfer data, often at HDRs, when needed.

The WUR is particularly good at addressing one of the most challenging power consumption problems, which is when the device needs to support low-latency data traffic while still maintaining low-power consumption. We can illustrate this problem with an example. Say the 802.11 device needs to control an actuator in a home automation use case. Let us say the response time for the actuator needs to be short since the user wants to visually see or hear the response. The user could be turning on a set of lights around the house, locking a set of doors, or maybe even remotely turning on some music. The user does not want to have to wait multiple seconds to see or hear the response to the action taken, say on a smartphone application, where the smartphone is connected to the 802.11 home network. The user wants to push a button on the smartphone and see or hear the response within a fraction of a second. To do this the smartphone needs to send a message over the

802.11 network to the AP, which then needs to send a message to the actuator over the same 802.11 network. The 802.11 device connected to the actuator needs to be ready to take action at any time, so it needs to be listening to the 802.11 network frequently, say every 100 ms, or possibly even more frequently. However, without a WUR, this means that the main receiver needs to listen to the AP very frequently and cannot spend much of its time in sleep mode. However, with a WUR the MR attached to the actuator can be in a deep sleep mode almost all the time, and the WUR can signal it to "wake-up" out of deep sleep mode, when an action is required. So we see how the ultra-low-power WUR can help solve this most difficult problem of supporting both low latency and lower power consumption.

1.2 Overview

This book provides a description of the IEEE 802.11ba standard, which as mentioned above is an amendment to the overall IEEE 802.11 standard. The IEEE 802.11 standard has grown over the years as more and more amendments have been added to the standard. IEEE 802.11ba is one of the more recent amendments. Many of the 802.11 amendments are focused on increases in wireless data rates and an increase in the overall network throughput. Many technologies are used to increase the data rate and the network throughput, like increased bandwidth, multiple input multiple output (MIMO) technology, and improvements in protocol design to optimize efficiency. IEEE 802.11ba is unique in that it focuses on power savings in order to increase the battery life in battery-operated devices.

The amendments that provide higher data rates typically also result in increased power consumption. Increases in bandwidth lead to increases in radio frequency (RF) and analog circuit power consumption. Use of MIMO technology requires duplication of RF/analog and digital processing circuits since circuits are needed for each spatial stream. With the use of more powerful forward error correction (FEC) codes, like low-density parity check (LDPC) codes, improved network performance comes at the cost of increased power consumption. Some of this is offset by improved integrated circuit technology, particularly for digital circuits. However, for RF and analog circuits, the improvements in integrated circuit technology do not typically result in significant offsets in power consumption.

The IEEE 802.11ba standard addresses these issues by providing a specification that enables implementations that require very low-power consumption. The standard does not mandate a specific implementation but makes it possible for an implementer to design an implementation with low-power consumption.

Details of the standard will be left for subsequent chapters, but here we can highlight some of the key aspects of the standard which enable a very low-power implementation.

First, let us explain that the standard was developed to enable an ultra-low-power wake-up receiver. The power consumption of the transmitter is not a major factor in the design. The reason for this is that in typical use cases the 802.11ba transmitter is implemented in the 802.11 AP, which is typically connected to AC power and is not battery-operated. On the other hand, the 802.11 station (STA) is often battery-operated and the 802.11ba wake-up receiver is implemented in this battery-operated device. So, the majority of design aspects of the 802.11ba standard are focused on saving power at the battery-operated STA.

The first design aspect of the standard that is different than other amendments is that the bandwidth is for 4 MHz operation, which is quite different than the amendments which focus on higher data rates, which have increased the bandwidth from 20 to 40, 80, and 160 MHz. There is also an amendment under development (IEEE 802.11be) which will support 320 MHz operation.

The second design aspect is the use of a simple on-off keying (OOK) modulation, which allows for the use of a non-coherent wake-up receiver. Most standards use modulations that require coherent receivers, leading to higher performance at the cost of higher power consumption at the receiver. The OOK modulation used in 802.11ba is a little different than traditional OOK, in that the underlying waveform is a multicarrier waveform. However, it is still possible to use a non-coherent receiver with this modulation. More details of the modulation will be provided in subsequent chapters.

Another design aspect to reduce power consumption is to use a low data rate (LDR). This enables an RF implementation with a higher receiver noise figure. The receiver noise figure is a measure of how much noise the receiver introduces to the received signal. A lower receiver noise figure means lower noise and better performance. But to obtain a lower noise figure usually requires higher power consumption. An RF low-noise amplifier (LNA) with a low noise figure requires a high operating current, which leads to increases in LNA power consumption. The power budget of the wake-up receiver implementation cannot support a high-power LNA, so the system design needs to address the higher receiver noise figure in some way.

By using a low data rate, the receiver can be built with a higher noise figure while still operating at a reasonable receive power level. By using a lower data rate, even with a higher receiver noise figure, the 802.11ba system can have a similar range as the main 802.11 radio. This is important so that when the main 802.11 radio is placed in a deep sleep mode, and the wireless link relies on 802.11ba, the supported range between the AP and the STA is not reduced.

All of these aspects of 802.11ba allow for ultra-low-power receiver implementations by saving power in the RF, analog, and digital circuits.

The impact of having an LDR puts strong requirements on the medium access control (MAC) layer design to fit the necessary information for the MAC messages into a small number of octets (bytes). In that way the overall time duration of the

802.11ba packets is not excessive. If the duration of these packets were too long, they would lead to an overall drop in network throughput, by using too much airtime. IEEE 802.11 refers to the airwaves as the wireless medium, so we say the medium utilization of the WUR packets cannot be too high, since that would prevent the other higher data rate 802.11 devices from using the wireless medium, which would result in an overall reduction in the wireless network throughput.

To avoid this problem, it is very important to design the MAC frames to be very short. The design of these MAC frames is described in detail in the subsequent chapters.

Another MAC layer design that is critical to power saving is to support duty cycling of the 802.11ba wake-up receiver. If the receiver can operate at a low-duty cycle, then the power consumption can be reduced even further. To support duty cycling of the 802.11ba receiver, a WUR beacon is used so that the 802.11ba receiver can maintain synchronization with the 802.11ba transmitter. The local clock within the 802.11ba receiver will drift between these WUR beacons, but the receiver can utilize the beacon to resynchronize its local clock with the clock in the 802.11ba transmitter. There is also a MAC layer procedure for establishing how frequently the beacon is transmitted and how often the 802.11ba wake-up receiver should be listening for messages, like a WUR beacon or a wake-up message. It is quite possible to duty cycle the 802.11ba receiver at 1% or less, leading to significant additional power savings.

So, in summary the following design aspects of the 802.11 standard enable ultra-low-power receiver implementations:

- Narrow bandwidth: Reduced RF and analog circuit power consumption.
- Non-coherent modulation: Enabling the use of a low-power non-coherent wake-up receiver.
- LDR: Enable a high receiver noise figure to save RF power consumption.
- Duty cycling of the 802.11ba wake-up receiver: Allowing the 802.11ba receiver to be in the powered-off mode for a high percentage of the time.

1.3 Book Outline

Chapter 2 provides a brief overview of the IEEE 802.11 standard, to provide the necessary background for the reader to understand the subsequent chapters on the WUR physical (PHY) and MAC layers. The focus is on the aspects of the standard that are relevant to the 802.11ba amendment. The chapter gives an overview of the orthogonal frequency division multiplexing (OFDM) PHY layer with a focus on the preamble design. There is also an overview of the IEEE 802.11 MAC layer, so the reader has a background on 802.11 MAC frames, procedures, and protocols.

Chapter 3 describes the concept of the WUR and how it is used to save power in the battery-operated STA. It includes a description of the primary circuits in the receiver, which consume significant amounts of power. It shows how a WUR can be used to save power, particularly in cases where the data transfer between the AP and the STA is infrequent. A description of the impact of latency between the AP and the STA is provided and how that impacts power consumption. The reader can then understand how, with the use of a WUR, it is possible to maintain a low latency link between the AP and the STA, along with low-power consumption at STA. This is one of the key benefits of using a WUR, which makes it possible to support both *low-latency operation* and *ultra-low-power consumption*. Without a WUR, that is very difficult to accomplish.

A description of the 802.11ba PHY layer is provided in Chapter 4. The PHY protocol data unit (PPDU) is described. There is an explanation of how the first portion of the 802.11ba PPDU is intended to be decoded by 802.11 devices that do support 802.11ba. These devices are often referred to as legacy devices since they do not understand the new 802.11ba PPDU. This is important so that those devices do not transmit at the same time as the 802.11ba PPDU, so as to avoid a packet collision. The chapter describes how the second part of the PPDU is intended to be decoded by the 802.11ba receiver. That portion consists of a synchronization (Sync) field which is used by the receiver for packet detection, timing recovery, and data rate determination. There are two data rates supported in the 802.11ba PHY layer: LDR and HDR. After the Sync field is the Data field, which carries the information provided by the MAC layer.

Chapter 5 focuses on the performance of the 802.11ba PHY layer. Simulation results are provided for different channel conditions and different signal-to-noise ratios (SNRs). The performance of packet detection by using the Sync field is provided for both the LDR and the HDR. Other simulations of Sync field processing included simulations of timing recovery accuracy and data rate determination. Finally, overall packet error rate (PER) is shown for additive white Gaussian noise (AWGN) channels and multipath channel models commonly used in 802.11 simulations. Finally, PER simulations are provided for the case when multiple transmit antennas are available at the AP, showing how the reliability of the link can be improved using multiple transmit antennas at the AP, which are typically available in most APs.

Chapter 6 gives an overview of the 802.11ba MAC protocol including the procedures for setting up a WUR link, setting up duty cycle operation, and how the WUR beacon works. This chapter describes how the WUR fits into the overall 802.11 power management system.

Chapter 7 gives a detailed description of the 802.11ba MAC frames, including the WUR Beacon frame, the Wake-up frame, the Discovery frame, and the Vendor Specific frame. The WUR Beacon frame is sent by the AP periodically so that the

802.11ba devices can maintain synchronization with the AP. This synchronization is important to support duty cycling of the 802.11ba client device. Without this synchronization the clocks in the AP and the client device would diverge and duty cycle operation would fail. The Wake-up frame can be used to wake up a single 802.11 client device or a group of 802.11 devices. The Discovery frame can be used to "discover" other 802.11ba devices which are within range, and that can be used to facilitate a variety of applications. One such application is when a mobile device wants to find an AP within its range. Using the 802.11ba Discovery frame, this discovery operation can be accomplished using an ultra-low-power wake-up receiver, resulting in lower power consumption. Finally, the Vendor Specific frame is an 802.11ba MAC frame that can be customized by a vendor which implements the 802.11ba standard for a custom use case. This provides great flexibility for use cases not foreseen during the development of the 802.11ba standard. Last but not least, Chapter 7 describes how existing management frames in 802.11 are expanded so that the 802.11ba WUR works seamlessly with the main 802.11 radio.

.

2

Overview of IEEE 802.11

2.1 Introduction

IEEE 802.11 is a wireless local area network (WLAN) standard. Here we provide a short overview of IEEE 802.11 as background for more detailed chapters on the IEEE 802.11ba wake-up radio standard. IEEE 802.11ba is an amendment to IEEE 802.11 adding new capabilities. There have been many amendments to the IEEE 802.11 standard over the years. The original IEEE 802.11 standard was published in 1997. Then in 1999 amendments 802.11a and 802.11b were published. Those amendments added higher data rate physical (PHY) modes. As one can see, these amendments are typically indicated by "802.11" with one or more letters appended. Some of the more famous amendments add major new PHY capabilities: 802.11a, 802.11b, 802.11n, 802.11ac, and most recently 802.11ax. There are many other amendments, some of which specify operation in millimeter wave frequency bands, like 802.11ad and 802.11ay. There are many other less well-known amendments that add additional capabilities, often in the medium access control (MAC) layer.

Here we will provide some background on PHY and MAC layers of IEEE 802.11. This background material is focused on the key aspects of the PHY and MAC layer that are important to understand 802.11ba and do not provide a detailed description of 802.11. For a more detailed description, the reader is directed to [1] which provides a detailed description of IEEE 802.11 with a focus on 802.11n and 802.11ac.

Section 2.2 gives an overview of the 802.11 PHY layer, with a focus on those aspects relevant to 802.11ba. In Section 2.3, we give an overview of the 802.11 MAC layer, with a focus on those aspects relevant to 802.11ba.

IEEE 802.11ba: Ultra-Low Power Wake-up Radio Standard, First Edition.
Steve Shellhammer, Alfred Asterjadhi, and Yanjun Sun.
© 2023 The Institute of Electrical and Electronics Engineers, Inc.
Published 2023 by John Wiley & Sons, Inc.

2.2   Overview of the IEEE 802.11 PHY Layer

The PHY layer specifies the details of transmission over the wireless medium. These specifications include things like the operating frequency, operating bandwidth, modulation, forward error correction (FEC) coding, and the like. Much of the PHY layer specification is on the PHY protocol data unit (PPDU), which can be thought of as a PHY packet. The primary purpose of the specification is to allow interoperability between different implementations. This means that if one company builds an implementation and another company builds a second implementation, then the two implementations can work together. In other words, if one implementation transmits a PPDU, then the other implementation can receive and process the PPDU. This requires a very detailed specification of the PHY layer in the IEEE standard.

2.2.1   Operating Frequencies and Bandwidths

One of the important parts of the PHY specification is the operating frequency of the standard or amendment. The 802.11 standard specifies operation in frequency bands that can be used without requiring a license. These frequency bands are often referred to as unlicensed or licensed-exempt frequency bands. These various frequency bands are regulated by local government agencies, like the Federal Communication Commission (FCC) in the United States. So even though a license is not required to operate in those bands there are regulations on transmission specifications like transmit power and bandwidth. These regulations are considered when developing new amendments to the 802.11 standard, so that implementations can be built that can operate in regions where these unlicensed frequency bands are available, typically all around the world.

Two of the more popular frequency bands used by 802.11 are the 2.4 and 5 GHz frequency bands. These frequency bands are widely available and used by some of the most popular 802.11 amendments like 802.11a, 802.11n, 802.11ac, and recently 802.11ax. These are the frequency bands specified in the 802.11ba amendment. One of the reasons that these frequency bands are specified in 802.11ba is that it allows the 802.11ba PHY to operate in a common frequency band with these popular amendments, which even allows for implementations that can share antennas, saving implementation costs. However, in an implementation 802.11ba is permitted to operate in a different frequency band than an implementation of another 802.11 amendment in the same device.

The regulations of these frequency bands were considered during the development of the 802.11ba amendment, which will be covered in Chapter 4.

Another fundamental aspect of the PHY is the channel bandwidth. As 802.11 has evolved, more channel bandwidths have been supported. Here we will focus on the orthogonal frequency division multiplexing (OFDM) PHY designs and not the earlier PHY designs before OFDM was introduced. In Section 2.2.2, we will provide a short overview of OFDM. In 802.11a the channel bandwidth supported is 20 MHz. The actual signal bandwidth is less than the channel bandwidth, so as to not interfere with the adjacent channels. Typically for a 20 MHz channel bandwidth the signal bandwidth is around 16–17 MHz depending on how it is measured. The standard specifies a spectral mask limiting how much power can be transmitted in a given portion of the spectrum. In 802.11n channel bandwidths of 40 MHz were introduced. As expected, the primary benefit of increased bandwidth is higher data rates, so when sufficient spectrum is available the increased bandwidth can be very beneficial. In 802.11ac channel bandwidths of 80 and 160 MHz were introduced providing increases in bandwidth when the spectrum is available. The 5 GHz frequency band allows for these wider channel bandwidths in many locations around the world.

2.2.2 OFDM

IEEE 802.11a introduced the OFDM PHY. OFDM stands for orthogonal frequency division multiplexing. OFDM is a multicarrier technique in which there are a number of subcarriers that are each modulated with BPSK, QPSK, or QAM. There is a specified frequency separation between these subcarriers and also a specified number of subcarriers. We will use 802.11a as an example since it was the first 802.11 OFDM PHY and also since some of the OFDM parameters in 802.11a apply directly to 802.11ba.

The subcarrier spacing is 312.5 kHz and there are a total of 64 subcarriers, which means if all the subcarriers were modulated, the signal bandwidth would be around 20 MHz (64×312.5 kHz = 20 MHz). However, some of the subcarriers are not modulated and have no energy placed on them. These subcarriers are often referred to as null subcarriers or null tones. For example, in 802.11a only 52 subcarriers are modulated, so the actual bandwidth is around 16.7 and not 20 MHz.

The OFDM transmitter typically uses an inverse fast Fourier transform (IFFT) to convert the frequency-domain subcarriers to a time-domain waveform. The IFFT will generate a time-domain waveform whose duration is the inverse of the subcarrier spacing, which is 3.2 μs in this case ($1/(312.5$ kHz$) = 3.2$ μs). This is sometime referred to as the IFFF duration. This time-domain waveform is preceded by a guard interval (GI) which is also often referred to as the cyclic prefix. The cyclic prefix is a copy of the tail end of the IFFT output placed before the IFFT output, to cyclically extend the waveform. A common duration of the cycle prefix is one-fourth of the IFFF output. So, in the case of 802.11a, the cycle prefix is a copy of the last fourth of the IFFT output, as shown in Figure 2.1. Here we see that the IFFT output is 3.2 μs in duration and so the last fourth is 0.8 μs in duration. That portion is copied and placed before the IFFT output to obtain a total OFDM

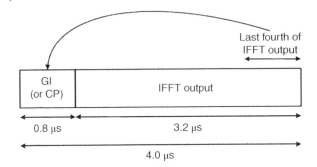

Last fourth of
IFFT output

GI (or CP)	IFFT output

0.8 µs   3.2 µs

4.0 µs

Figure 2.1 OFDM symbol with cyclic prefix.

symbol duration of 4.0 µs. There is optional support for a shorter GI, but since the 0.8 µs GI is also used in 802.11ba we will focus on this numerology.

The OFDM symbol is used to construct the PPDU which is the PHY-layer packet, which we describe in Section 2.2.3.

2.2.3 OFDM PPDU

The PPDU can be divided into the preamble and data sections. The preamble is designed to facilitate the receiving process by providing the structure to enable a number of steps of the receiving process: packet detection, coarse timing and fine timing recovery, coarse and fine frequency offset correction, and finally channel estimation. We will give a brief overview of the 802.11a OFDM preamble and then give a brief description of how this preamble has been extended in subsequent OFDM PHY amendments. We will not go into the details of the data portion of the PPDU since the 802.11ba amendment is unique in how it modulates the signal in the Data field. In 802.11ba, the OFDM preamble portion is decoded by non-wake-up radio (non-WUR) stations (STAs) so that they are aware of the presence of a PPDU and the duration of the PPDU, so that those STAs do not transmit during the 802.11ba PPDU. So, some understanding of the OFDM preamble is useful for understanding the 802.11ba preamble.

The 802.11a PPDU structure Is shown in Figure 2.2. Details are available in Clause 17 of the IEEE 802.11 standard [2].

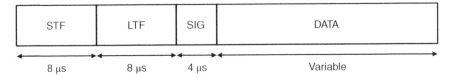

STF	LTF	SIG	DATA

8 µs   8 µs   4 µs   Variable

Figure 2.2 IEEE 802.11a PPDU structure.

The 802.11a preamble consists of the short training fields (STF), the long training fields (LTF), and the signal (SIG) field.

The STFs are constructed with only every fourth subcarrier populated with a modulated signal. This is useful since there is likely a frequency offset between the transmitting STA and the receiving STA, and with this design the receiving STA can perform coarse frequency offset correction. There are a total of 10 STFs in the STF for a duration of 8 μs. The LTF consists of a pair of LTFs, each 3.2 μs in duration preceded by a long guard interval of duration 1.6 μs, for a total LTF duration of 8 μs.

The receiving STA uses the STF for packet detection, automatic gain control (AGC), coarse timing recovery, and coarse frequency offset correction. Then the receiving STA uses the LTF for fine timing, frequency correction, and channel estimation.

After the LTF, there is the SIG field which carries the information needed for decoding the Data field. This information is encoded using binary phase shift keying (BPSK) and a rate 1/2 convolutional FEC coding. The data rate of SIG is 6 Mb/s. We will not cover all the details of the SIG field contents, but we will highlight two of the subfields: RATE and LENGTH. The RATE subfield indicates the data rate used in the DATA field. There are several rates available starting at 6 Mb/s and increasing up to 54 Mb/s. The second subfield of note is the LENGTH subfield which signals the number of octets (bytes) in the PHY service data unit (PSDU), which is the information content sent in the Data field. From the data rate and the number of octets in the PSDU, the receiving STA can calculate the number of OFDM symbols in the DATA field.

The next OFDM PHY that was developed is 802.11n, which we will briefly describe here. Details of 802.11n can be found in Clause 19 of the IEEE 802.11 standard [2]. The 802.11n PPDU structure is shown in Figure 2.3.

The STF, LTF, and SIG fields from the 802.11a PHY were relabeled as legacy fields and referred to going forward as L-STF, L-LTF, and L-SIG. The 80211n PHY is referred to as the high-throughput (HT) PHY, so some of the new fields use the "HT" prefix. The original 802.11a PHY is then reclassified as the non-high-throughput (non-HT) PHY.

The L-SIG is followed by the 8 μs HT-SIG field, which carries the information about the structure of the DATA field. An interesting aspect of the HT-SIG is that it uses a unique modulation referred to as quadrature binary phase shift keying

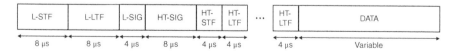

Figure 2.3 IEEE 802.11n PPDU structure.

(Q-BPSK) where the BPSK constellation is along the quadrature axis (*y*-axis) versus the in-phase axis (*x*-axis). So, the Q-BPSK constellation is rotated 90° relative to the BPSK modulation used in the L-SIG. This allows the receiving STA to distinguish between an 802.11a non-HT PPDU and an 802.11n HT PPDU. This PPDU classification process is often referred to as autodetection.

The HT-SIG is followed by a 4 μs HT-STF which can be used by the receiving STA to perform additional AGC, which is useful in the cases where beamforming has been applied to the PPDU after the legacy fields, since the beamforming typically changes the receive power level at the receiving STA. This is followed by several HT-LTFs. More than one is needed since 802.11n introduced multiple input multiple output (MIMO) technology into IEEE 802.11. There are up to four MIMO spatial streams supported in 802.11n. There are more HT-LTFs needed when MIMO is used, the details of which are beyond the scope of this brief overview.

The next major PHY after 802.11n was 802.11ac, which is also known as the very high throughput (VHT) PHY. Like before, the new PHY introduced new capabilities like support for wider bandwidth and other such features providing an increase in throughput. As before, we will focus on the preamble, since that is the portion that has an impact on the 802.11ba design.

Once again, a detailed description of 802.11ac is available in [1] as well as in Clause 21 of the IEEE 802.11 standard [2]. The 802.11ac PPDU structure is shown in Figure 2.4.

As previously the preamble is designed so the receiving STA can autodetect between VHT, HE, and non-HT, so that the receiver knows how to process the remainder of the PPDU.

The VHT preamble begins with the legacy short training fields (L-STF), legacy long training fields (L-LTF), and the legacy signal field (L-SIG), just like in the HT PHY. That is followed by the very high throughput signal field A (VHT-SIG-A). The duration of this field is 8 μs just like the HT-SIG in 802.11n.

Recall that HT-SIG uses Q-BPSK to enable the receiving STA to differentiate it from non-HT.

Here the phase rotation approach is modified so that the receiving VHT STA can distinguish the VHT preamble from both the non-HT and HT preambles.

The 8 μs VHT-SIG-A field consists of two 4 μs OFDM symbols, corresponding to subfields VHT-SIG-A1 and VHT-SIG-A2. The first subfield, VHT-SIG-A1, uses

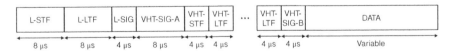

Figure 2.4　IEEE 802.11ac PPDU structure.

standard BPSK and the second subfield, VHT-SIG-A2, uses the special Q-BPSK. So, the receiving VHT STA uses the modulation of those two symbols to differentiate between non-HT, HT, and VHT.

The VHT-SIG-A field is followed by the VHT-STF, a set of VHT-LTFs, a second signal field called the very high throughput signal field B (VHT-SIG-B), and finally, the Data field.

Finally, the most recent mainstream PHY is the 802.11ax PHY which is also referred to as the high efficiency (HE) PHY. Once again, this PHY incorporates additional features to increase throughput and also in this case, improve the overall network efficiency.

We will focus on the preamble design for autodetection since that is the portion that affects the 802.11ba PPDU design. Details of the HE PHY can be found in the high-efficiency amendment [3] to the 802.11 standard.

There are several HE PPDU types: single user (SU), multiuser (MU), extended range (ER), and trigger based (TB). We will use the SU PPDU type for our description of the autodetection technique. The same process applies to the other PPDU types.

The 802.11ax SU PPDU structure is shown in Figure 2.5.

The autodetection design for 802.11ax is different in nature than that of the previous PHYs, which relied on the use of BPSK or Q-BPSK in the two OFDM symbols after the L-SIG. The 802.11ax preamble uses BPSK for the two symbols after the L-SIG, so from a modulation point of view, it is the same as the non-HT PPDU. Another detail of L-SIG in these cases is that the legacy data rate that is set in the L-SIG for HT, VHT, and HE is 6 Mb/s which corresponds to using BPSK with a rate-1/2 convolutional code. With that data rate in non-HT, each OFDM symbol carries three octets, so in non-HT the length of the PSDU in octets is a multiple of three. In other words, the value of the length field modulo 3 is zero. So, we sometimes write mod(Length, 3) = 0.

In 802.11ax, the value of the length field is selected not to be a multiple of 3. In particular, for MU and ER PPDUs, the length field in the L-SIG is selected so that mod(Length, 3) = 1. And for the SU and TB PPDUs, the length field in the L-SIG is selected so that mod(Length, 3) = 2. Additionally, after the L-SIG, there is a repeated version of the L-SIG, referred to as the repeated legacy signal field (RL-SIG). The receiving HE STA can classify the PPDU as an HE PPDU by this repeated L-SIG. In addition, this can be done sooner than waiting for the second symbol after the L-SIG, which was required in HE and VHT PPDU classification.

Figure 2.5   IEEE 802.11ax SU PPDU structure.

In Chapter 4, we will look at the preamble design for 802.11ba, which is designed so that for STAs developed before 802.11ba will classify the PPDU as a non-HT PPDU. The details of that design are provided in Chapter 4.

For the PPDU designs, here we focused on the preamble design since that is what affects the portion of the 802.11ba PPDU preamble design which is to be decoded by non-802.11ba STAs, so that those STA defer to 802.11ba PPDUs and do not transit during the 802.11ba PPDUs. For more details on the 802.11 PHY designs, see [1–3].

2.3 Overview of IEEE 802.11 MAC Layer

The MAC layer is built on top of the PHY layer to provide many key network services including:

- Network discovery
- Connection setup
- Coordinated wireless medium access
- Security

The MAC layer also defines various mechanisms to improve network efficiency when providing the above services. The key mechanisms include but are not limited to:

- Time synchronization
- Power-saving mechanisms
- Orthogonal frequency division multiple access (OFDMA)

In the rest of this section, we will provide an overview of these services and related mechanisms as they are closely related to the 802.11ba MAC features.

2.3.1 Network Discovery

For network discovery, the MAC layer defines a passive scanning mechanism via Beacon frames and an active scanning mechanism via Probe Request/Response frames. These two mechanisms help a STA to discover an AP including its ID, capabilities, and operation parameters. With the discovered information, the STA may decide whether to set up a connection to the AP.

An AP typically transmits Beacon frames with a fixed beacon interval, whose default value is 100 TU, which is equivalent to 102.4 milliseconds (ms). The Beacon frames include the following information about the AP:

- Service set identifier (SSID), which is an identifier of the network.
- PHY and MAC capabilities, such as supported data rates.

- Operation parameters, such as the operating channel configuration.
- Timestamp, which is used to provide synchronization between the AP and the STA.

With the passive scanning mechanism, the STA listens to a channel to detect Beacon frame transmitted on that channel by AP(s) within its proximity. With the active scanning mechanism, the STA transmits a Probe Request frame requesting the APs within proximity to the STA to respond with a Probe Response frame that carries similar information which is carried in the Beacon frame. The Probe Response frame can be transmitted as soon as possible, but the Beacon frame can only be transmitted at the frequency determined by the beacon interval. Such flexibility in scheduling helps the active scanning mechanism achieve a shorter discovery time than the passive scanning mechanism. The shorter discovery time, however, is at the cost of increased messaging overhead from the Probe Request/ Response frames.

Figure 2.6 illustrates how STA1 discovers an AP operating on Channel 36 using passive scanning and STA2 discovers the AP using active scanning. The AP transmits a Beacon frame every beacon interval. With passive scanning, each STA listens on Channel 36 passively until it receives a Beacon frame. To discover the AP using passive scanning, STA1 needs to be prepared to wait on the channel long enough to ensure the AP has at least transmitted a Beacon frame during the time in which the STA waits on the channel. The discovery delay can be as long as one beacon interval, as it largely depends on how soon the AP will transmit the next Beacon frame. Active scanning can be used to shorten the discovery delay, as illustrated using STA2. In this example, STA2 transmits a broadcast Probe Request frame as soon as the medium becomes idle on Channel 36, soliciting a response from any AP on that channel. Upon receiving the Probe Request frame, the AP can typically respond with a Probe Response frame quickly if the medium is idle. The Probe Response frame carries almost identical information as a Beacon frame

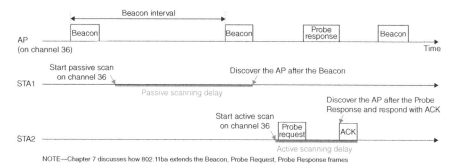

Figure 2.6 Network discovery with passive scanning or active scanning.

and enables STA2 to discover the AP with much shorter discovery delay. However, such shortened delay is at the cost of increased medium utilization. As illustrated in the figure, there are three extra frames exchanged over the air, the Probe Request frame from STA2, the Probe Response frame from the AP, and the acknowledgment (ACK) frame from STA2 in response to the Probe Response frame.

In practice, it is up to a STA to determine whether to use the passive scanning mechanism or the active scanning mechanism, depending on whether the STA prefers to minimize the discovery latency or the messaging overhead. In order to support various use cases, 802.11ba has been designed to work with both the passive scanning mechanism and the active scanning mechanism, by enhancing the Beacon frame, the Association Request frame, and the Association Response frame. As the existing scanning mechanisms are energy-consuming and time-consuming, 802.11ba has also defined an alternative mechanism to discover an AP via the 802.11ba PHY with much lower energy consumption. All these enhancements will be discussed in Chapter 6.

2.3.2 Connection Setup

After discovering an AP that a STA prefers, to gain full access of the AP, the STA sets up a connection with the AP based on Association Request/Response frames. The key objective of the Association Request/Response frames is to reach an agreement between the AP and the STA on many key operating parameters such as the minimum set of data rates that need to be supported.

A typical connection setup is illustrated in Figure 2.7. In this example, A connection setup is initiated by the STA by transmitting an Authentication Request frame to the AP. If the AP is willing to accept the request, the AP responds to the STA with an Authentication Response frame. Although not illustrated in the figure, the AP may refuse the request in some situations, for example, when the AP is

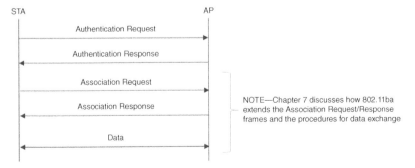

Figure 2.7 Connection setup.

overloaded. More details of the Authentication Request/Response frames are available in [1]. Next, the STA transmits an Association Request frame to the AP. The Association Request frame includes the PHY/MAC capabilities of the STA, which help the AP find out how to best communicate with the STA. As a response to the Association Request frame, the AP decides to accept or refuse the association request by an Association Response frame. The Association Response frame also assigns an association ID (AID) for the STA to use in some 802.11 mechanisms such as the power save (PS) mechanism discussed in Chapter 6.

After the successful association, the STA and AP are ready to exchange data.

2.3.3 Coordinated Wireless Medium Access

To coordinate STAs in an 802.11 network to efficiently share the same wireless medium, the 802.11 standards have defined multiple coordination functions. The primary one is called the distributed coordination function (DCF), which is also known as carrier sense multiple access with collision avoidance (CSMA/CA).

With DCF, before a STA attempts to transmit on the wireless medium, the STA first senses the medium to detect whether any other STA is currently transmitting. This is referred to as physical carrier sense. If the medium is determined to be idle, then the STA starts its transmission. Otherwise, the STA defers its transmission until the medium becomes idle. The physical carrier sense mechanism mentioned above is typically based on the detection of energy or detection of the PPDU preamble on the wireless medium. Such carrier sense operation helps the STA to avoid collisions with ongoing transmissions over the wireless medium.

The 802.11 amendments are always designed with backward compatibility in mind, including the physical carrier sense. This is because the STAs from different 802.11 amendments often need to coexist with each other on the same operating channel. For example, STA1 may be a very old laptop that only supports 802.11a, whereas STA2 in the same network may be a recent smartphone that supports 802.11ax. For physical carrier sensing to work well between the two STAs, a PPDU defined in newer amendments often begin with a legacy PHY preamble, noted as L-PHY in Figure 2.8. The L-PHY carries information on the duration of the PPDU and is understandable by all 802.11 STAs.

Figure 2.8 L-PHY for better coexistence with legacy STAs.

In this example, the 802.11ax STA2 can decode the 802.11a PPDU from the 802.11a STA1, because 802.11ax is built upon 802.11a. However, STA1 is not able to decode the whole 802.11ax PPDU from STA2 as STA1 is only designed to decode the legacy PPDUs defined in 802.11a. Thanks to the L-PHY prefix of the 11ax PPDU, however, STA1 can still detect the presence and duration of the 802.11ax PPDU for correct physical carrier sense. As discussed in detail in Chapter 6, 802.11ba also prepends an L-PHY to 802.11ba frames to ensure backward compatibility.

2.3.4 Enhanced Distributed Channel Access

Collision avoidance is introduced because carrier sense cannot prevent packet collisions all the time. For example, two STAs may both sense an idle wireless medium and start their transmissions at exactly the same time, resulting in a packet collision. To prevent the STAs from repeated collisions like this, DCF requires each STA to perform a random backoff before the next transmission attempt. The random backoff helps to create a random offset between the next transmission attempts from the STAs, which in turn helps to prevent repeated collisions between the STAs.

In addition to DCF, 802.11 has also defined a medium access mechanism called enhanced distributed channel access (EDCA). EDCA provides distributed and differentiated access to the wireless medium, which enables an 802.11 network to provide prioritized access for time-sensitive traffic such as voice and video.

These EDCA/DCF mechanisms are typically used in an 802.11 network for all frame types including Data, Management, and Control frames. For better coexistence with legacy STAs, new frames introduced in 802.11ba (see Chapters 6 and 7) also need to follow these EDCA/DCF mechanisms.

2.3.5 Security

To secure 802.11 transmissions, various mechanisms have been defined for data confidentiality and integrity in an 802.11 network, such as the counter mode with cipher-block chaining message authentication code protocol (CCMP) and the galois/counter mode protocol (GCMP). The core idea behind these mechanisms is to help a pair of STAs to mutually verify each other's identities and to generate dynamic encryption keys for secure data transmissions between them. These mechanisms prevent an attacker from impacting the privacy of a user, the integrity of transmitted data over the air, or the availability of the network.

Here we focus on the security enhancements specific for 802.11ba. Chapter 6 describes how these security mechanisms have been expanded to protect the extended frames or newly defined frames in 802.11ba. Considering the unique characteristics of the 802.11ba PHY, the security expansion in 802.11ba has also defined new security mechanisms that are specifically optimized to conserve power at a battery-powered STA.

2.3.6 Time Synchronization

An AP and STAs associated with the AP are synchronized to a common clock using a timing synchronization function (TSF). The AP serves as the timing master for the TSF and conveys the TSF to the STAs in each Beacon frame as a timestamp of 8 octets. The timestamp indicates the number of microseconds since the start of the AP. Upon receiving the Beacon frame, the STAs update their local TSFs based on the timestamp.

The common clock is used for many procedures in an 802.11 network. For example, the AP and STAs use the TSF to determine the schedule to transmit or receive Beacon frames, respectively. At every target beacon transmission time (TBTT), the AP will schedule a Beacon frame transmission. Time zero is defined to be a TBTT and the successive TBTTs are derived based on the beacon interval discussed in Figure 2.6. In case the beacon interval is set to 100 TU, the successive TBTTs will be at 100 TU, 200 TU, and so on. It is worth noting that a Beacon frame transmission may not start exactly at a TBTT due to other ongoing transmissions on the wireless medium as discussed in Section 2.3.4. To overcome such uncertainty, the AP does not update the timestamp in each Beacon frame right until its transmission can be started.

To minimize energy consumption at the STAs, 802.11ba has introduced duty cycling that is also based on the common clock, as discussed in detail in Chapter 6.

2.3.7 Power-Saving Mechanisms

IEEE 802.11 has been implemented in more and more battery-powered devices such as smartphones, wireless doorbells, and VR goggles. To extend the battery life of these devices, a set of PS mechanisms have been defined in 802.11 such as the legacy PS mode based on the Power Management bit in uplink frames, the unscheduled automatic power save delivery (U-APSD), and the target wake time (TWT).

The core idea behind the power management mechanisms is to put the STA's radio into a sleep/doze state as long as possible. This is because the radio typically consumes less power in the doze state compared to a receive/awake state, as discussed in detail in Chapter 3. The key limitation with the doze state is that the radio cannot receive incoming data in this state. To get around this limitation, these PS mechanisms have provided means for the STA to notify the AP of the power management state of the STA's radio, so that the AP only transmits to the STA when the radio is awake.

Figure 2.9 illustrates the state machine for power management. At the high level, a STA is either in the active mode or in the PS mode. In the active mode, the STA has its radio in the awake state, in which the STA is ready to receive incoming data. In the PS mode, the STA's radio is allowed to move between

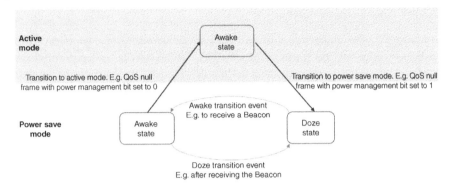

Figure 2.9 Power management.

the awake state and the doze state. With the radio in the doze state, the STA saves more power but cannot receive incoming data.

To avoid incoming data when the STA's radio is in the doze state, IEEE 802.11 has defined methods for the STA to notify the AP of the transition event to the doze state in the PS mode. In this example, the STA sends to the AP a frame with the Power Management bit set to 1 as the notification. The frame is typically a quality-of-service (QoS) Null frame or a data frame. Upon receiving the frame, the AP will buffer the data destined to the STA until the AP is notified that the STA's radio has moved to the awake state.

The STA learns that there is data for it buffered at the AP typically using one of two methods. First, the AP will indicate that there is buffered data in a Beacon frame with a traffic indication map (TIM) element. To learn the buffer data based on the TIM element, the STA needs to move its radio to the awake state and wait till the next TBTT to receive the Beacon frame. In this case, the STA remains in PS mode. If the received Beacon indicates no buffered data for the STA, the STA may move the radio back to the doze state. If there is buffered data, the STA may move to active mode to retrieve the data. If the STA wants to learn sooner if there is buffered data for it instead of waiting for the next Beacon frame, the STA can also proactively notify AP that it has moved to the awake state of the active mode by transmitting a frame, such as a QoS Null frame, to the AP with the Power Management bit in the frame set to 0. Upon receiving the frame, the AP can start transmitting the buffered data to the STA immediately.

Although the existing PS mechanisms help to reduce power consumption at the STA substantially, there is still room for improvement. In case there is no buffered data at the AP, it is a total waste for the STA to turn on its radio either for receiving the Beacon frame or for transmitting the QoS Null frame. Both methods given in the example above are energy-consuming and time-consuming to detect buffered data, because the STA needs to turn on all components of its radio, as discussed in detail in Chapter 3. To address such deficiency, 802.11ba has defined new capabilities in both PHY and MAC, as described in the remaining chapters.

2.3.8 Orthogonal Frequency Division Multiple Access (OFDMA)

IEEE 802.11ax has introduced OFDMA that allows an AP to simultaneously communicate with multiple STAs by dividing the operating channel into smaller subchannels named resource units (RUs). Compared to legacy non-OFDMA, OFDMA is more efficient in throughput and latency, especially when there are many contending STAs, each with limited amount of data. Example use cases include the 802.11 networks at airports or coffee shops.

Figure 2.10 illustrates the difference between legacy non-OFDMA (the PPDU transmissions defined prior to 802.11ax) and OFDMA defined in 802.11ax. In legacy non-OFDMA, data destined to one or more STAs will occupy the whole operating bandwidth of the AP. It is up to the AP to decide whether to transmit to one STA or multiple STAs. Assume that the AP has small amount of buffered data for each of the multiple STAs. If the AP chooses to transmit only to one STA in the non-OFDMA PPDU, other STAs likely experience higher latency compared to the chosen STA, as they have to wait for their data in a future PPDU. If the AP wants to transmit to multiple STAs in the non-OFDMA PPDU using multiuser multiple input multiple output (MU-MIMO), a far-away STA may have to be left out, as MU-MIMO typically requires the participating STAs have sufficient signal to interference plus noise ratio (SINR).

To avoid these limitations associated with the legacy non-OFDMA PPDU, the AP can divide the whole bandwidth into multiple RUs to serve multiple devices simultaneously using OFDMA. In Figure 2.10, the AP divides the whole bandwidth into 3 RUs to serve 3 STAs in one OFDMA PPDU. The AP may allocate a larger RU to a STA which has more buffered data at the AP, such as the STA2 in this example. In summary, OFDMA provides the AP a way to serve multiple STAs efficiently within the same bandwidth.

In Section 6.5, this book will discuss how 802.11ba leverages a feature similar to OFDMA to wake up multiple STAs that have buffered data at the AP in order to reduce the latency for the STAs to retrieve their data.

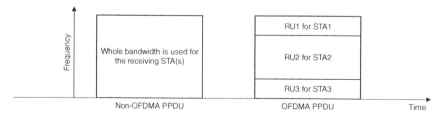

Figure 2.10   Illustration of OFDMA introduced in 802.11ax.

2.4   Conclusions

IEEE 802.11 has defined many more services and functionalities than what have been discussed here. Instead of a thorough overview of 802.11, our focus in this chapter is to cover the key aspects of the PHY and MAC layer that are important to understand 802.11ba discussed in the remainder of the book. As indicated multiple times in this chapter, many 802.11ba features utilize existing 802.11 services or procedures. The key takeaway is that 802.11ba is not a standalone feature. Instead, it has been designed to enhance and to work smoothly with existing 802.11 amendments.

References

1 Perahia, E. and Stacey, R. (2013). *Next Generation Wireless LANs: 802.11n and 802.11ac*. Cambridge University Press.

2 IEEE 802.11-2020 (2020). *IEEE Standard for Local and Metropolitan Area Networks: Wireless LAN Medium Access Control (MAC) and Physical Layer (PHY) Specifications*. IEEE.

3 IEEE 802.11ax (2021). *IEEE Standard for Local and Metropolitan Area Networks: Wireless LAN Medium Access Control (MAC) and Physical Layer (PHY) Specifications: Amendment: Enhancements for High Efficiency WLAN*. IEEE.

3

Wake-up Radio Concept

3.1 Introduction

The purpose of a wake-up radio (WUR) is to save power consumption in an IEEE 802.11 station. In this chapter we describe some of the primary sources of power consumption and how the WUR can be used to reduce power consumption under certain conditions.

In Section 3.2 we describe the various sources of power consumption in an IEEE 802.11 station. To understand those power consumption sources, we need to consider different operating modes, like transmit, receive, sleep, etc. We also need to understand and discuss the basic blocks operating in these modes, including both RF, analog, and digital circuits. We will not go into detail on specific power consumption values but describe the overall picture and what circuits typically dominate the power consumption in the various operating modes.

In Section 3.3 we describe the WUR concept and how it can be useful under certain conditions in saving a significant amount of power consumption at the IEEE 802.11 station. As we understand in what conditions we can save power consumption, it allows us to think about various use cases where a WUR can be beneficial.

In Section 3.4 we give a numerical example showing the power consumption using a WUR. Here, we use example power consumption values which might be considered reasonable but are not for any specific implementation. However, the methodology for performing the power consumption can easily be applied to a specific implementation, given the proper power consumption values of the various operating modes.

IEEE 802.11ba: Ultra-Low Power Wake-up Radio Standard, First Edition.
Steve Shellhammer, Alfred Asterjadhi, and Yanjun Sun.
© 2023 The Institute of Electrical and Electronics Engineers, Inc.
Published 2023 by John Wiley & Sons, Inc.

3.2 Primary Sources of Power Consumption in an IEEE 802.11 Station

Here we discuss the primary sources of power consumption in an IEEE 802.11 station. We typically consider the non-AP station, since in the majority of use cases, the access point is operating of AC power while in the cases of interest the non-AP station is operating using battery power. The lower the average power consumption the longer the battery life for a given battery size. Or another way of saying it, if the power consumption is reduced significantly then a smaller, lower capacity, battery could be used.

Let us first consider the various operating modes of an IEEE 802.11 station. The most obvious ones are transmit and receive. The station transmits packets to the AP either to carry data or to carry control/management frames. And of course, the station receives packets from the AP to receive data or control and management frames. One of the common packets for the station to receive is a packet carrying a beacon frame. These beacon frames are sent by the AP on a periodic basis, typically every 100 ms.

Other modes which are a little less obvious are sleep modes. Most implementations of non-AP stations support one or more sleep modes. During these modes many of the circuits are in some form of standby mode or potentially turned off. We will give examples of two generic sleep modes, which we will call "sleep" and "deep sleep." Different implementations may use different terminology and may support fewer or more sleep modes.

3.2.1 Power Consumption in Transmit Mode

There are many functions that consume power in the transmit mode. Here we consider a generic transmitter and do not go into details about each of the various circuits. The goal is to give a high-level overview and to indicate the primary power consumers. There are many functions in a transmitter. Here we will focus on the ones most important in terms of power consumption. The process begins at the upper layers, like transmit control protocol/internet protocol (TCP/IP) which typically runs in software. Then there is the medium access control (MAC) layer, which builds the MAC protocol data unit (MPDU). There are a number of steps in this process, and this also typically occurs in software. The power consumption of these steps is based on the processor which is being used and the clock rate, and other factors that affect the processor power consumption. Clearly, the higher the data rate the higher the power consumption, in general. Some function like calculating the cyclic redundancy code (CRC) may also be built in hardware given the strict timing requirements in 802.11. There is also memory that is used by the processor to store both code and data, which can also consume a

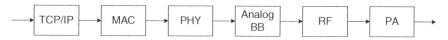

Figure 3.1 Generic transmitter block diagram.

significant amount of power. The MAC layer sends the MPDU to the PHY layer which constructs the PHY protocol data unit (PPDU). This layer can consume more power than the MAC layer since some of the operations can be more complex (Figure 3.1).

The PHY generates the preamble, which sends important information to the receiver so it can process the Data field. Most 802.11 versions use orthogonal frequency division modulation (OFDM), which requires an inverse fast Fourier transform (IFFT) to convert from the frequency domain to the time domain prior to the IFFT there is forward error correction (FEC) coding block (either a binary convolutional code (BCC) or a low-density parity check (LDPC) code). The FEC encoder is typically significantly less complex than the decoder, which is used at the receiver, but it does consume some power. The bandwidth of the 802.11 standards have increased over the years from 20 up to 160 MHz and are expected to increase up to 320 MHz in a future version (802.11be). This increased bandwidth increases the data rate, but that also increases the clock rate of the circuits leading to an increase in power consumption. Another factor in the PHY, is the support for multiple input multiple output (MIMO) technology, which also increases the data rate, but as a result increases the power consumption. The 802.11 standard supports up to eight MIMO streams of data. These streams of data are sent to parallel analog baseband (BB) and radio frequency (RF) circuits which are ultimately sent over multiple transmit antennas.

The MAC and PHY are implemented in software and digital hardware. Often the main power consumer is the analog BB and RF circuits. This beings in the transmitter with a pair of mixed-signal digital-to-analog converters (DAC) which results in the analog in-phase and quadrature BB signals. These analog signals are converted up to an RF frequency using a pair of mixers, which mix the BB signals with an RF carrier signal. This RF carrier signal, often referred to as the local oscillator (LO) is typically generated by a frequency synthesizer, which is designed to generate different carrier frequencies, in order to support multiple RF channel frequencies. IEEE 802.11 supports RF channels in the 2.4 and 5 GHz frequencies, and with the new 802.11ax amendment, it also supports channels in the 6 GHz band. Finally, the modulated RF signal is amplified by a pre-amplifier and then a power amplifier (PA). The PA can be one of the largest power consumers, since it is the job of the PA to amplify the signal to a power level needed to transmit the power over a significant range. For OFDM the signal power level at the output of the PA must be reduced from the maximum supported PA output power in order to operate in the linear operating range of the PA. This is commonly referred to PA back-off. The PA back-off is typically larger when higher order modulation is used, to avoid distortions caused by nonlinear effects in the PA. This causes the power consumption of the PA to be significantly higher than the actual PA output power. This is because the power consumption of the PA is typically related to the maximum supported RF power level and not the output power level of the actual signal being transmitted.

Just like the PHY, there are a number of factors that affect increases in analog/RF power consumption, in more recent versions of the 802.11 standard. The increase in channel bandwidth increases the power consumption of the DAC and the analog BB circuits. The use of higher carrier frequencies like 5 and 6 GHz, compared to 2.4 GHz increases the operating frequency of the RF circuits, which causes an increase in power consumption. Finally, the transmission of multiple streams of data, causes duplication of many of the analog/RF circuits, which can significantly increase power consumption.

Based on these observations, for low-power 802.11 stations one would expect implementations in the 2.4 GHz frequency band with only a few spatial streams of data. Even in those cases, the transmit power consumption can be significant.

3.2.2 Power Consumption in Receive Mode

The other main operating mode is the receive mode. This is when the station receives a packet from the AP. The receiver processing begins with the RF/analog circuits, then proceeds to the PHY and MAC circuits (Figure 3.2).

The receiver does not require a PA, so its overall power consumption is typically lower than that of the transmitter. However, there are still a number of RF/analog circuits which can consume significant power. The first RF circuit is the low-noise amplifier (LNA). This circuit amplifies the receive signal while adding as little electronic noise as possible. One of the main characteristics of the LNA is its noise figure. This factor measures the decrease in the signal-to-noise ratio (SNR) due to the LNA, as measured in dB. To have a low noise figure requires spending significant power in the LNA. The LNAs in many 802.11 receivers are designed to have a low noise figure, which results in better SNR, ultimately resulting in better receiver sensitivity and hence better range, all other things being equal. After the LNA there may be additional amplifiers, but they typically consume less power

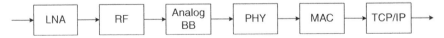

Figure 3.2 Generic receiver block diagram.

since after the receive signal has already been amplified, the effect of additional noise is less significant.

The RF signal is then converted down to BB frequency using a pair of mixers where the receive signal is mixed with the LO signal. The LO signal is generated by a frequency synthesizer, which may be the same frequency synthesizer used in the transmitter. The reason this circuit can be shared is that the station is never in transmit mode and receive mode at the same time. The in-phase and quadrature BB signals are filtered to remove noise and then sampled by a pair of analog-to-digital converters (ADCs). Then the digital BB signals are processed by the PHY layer.

Before moving on to discuss the PHY layer processing, we should mention the impact of increased channel bandwidth and MIMO. As we mentioned earlier when discussing the transmitter, the channel bandwidth has increased over the years from 20 up to 160 MHz today, and up to 320 MHz in the future. This does not impact the power consumption of the RF circuits too much, but it does increase the power consumption of the analog amplifiers and filter processing the in-phase and quadrature signals. It also increases the sampling rate of the dual ADCs. Finally, there may be multiple receiver antennas, and for each receive antenna, this chain of RF/analog circuits must be replicated. Clearly, this increases the power consumption of the receiver.

In lower power applications the station may stick with lower channel bandwidths of 20 or 40 MHz, and may have only one or two receive antennas. That is a very reasonable way to keep the receiver power consumption from being too high.

The PHY performs a number of functions on the digital BB signal. It must detect the presence of a packet. This is because there is not always a packet present and even when it is present the timing of the arrival of the packet can vary for many reasons. Once the packet has been detected, the PHY needs to perform a number of functions like identifying the timing of the packet arrival and compensating for any offset in the carrier frequencies between the transmitter and the receiver. Fields in the preamble are decoded to inform the receiver about the details of the packet, including the packet bandwidth, the modulation and coding scheme (MCS) used in the Data field, and so forth. Then an estimate of the channel on each OFDM subcarrier must be performed. After these functions are performed using the preamble the data must be processed using a fast Fourier transform (FFT) to demodulate the modulated signals on each OFDM subcarrier. Then those demodulated signals are regrouped based on the interleaver and then the FEC decoder decodes the data. The FEC code may be either a BCC or an LDPC code. The data decoded by the PHY is transferred to the MAC layer for further processing. As in the transmitter, the higher the bandwidth the higher the required clock rate of the PHY circuits leading to increased power consumption. Also, the higher data rates, using higher order modulation, increases the processing of the receiver circuits. And as mentioned before, if multiple streams of MIMO data are sent, this leads to the replication of many of the PHY circuits, leading to higher power consumption.

The MAC processes the data from the PHY. If has become common for the MAC to combine multiple MAC protocol data units (MPDUs) into an aggregated MAC protocol data unit (A-MPDU). At the receive the MAC must parse the A-MPDU which requires searching for delimiters and such. This is often performed in software that consumes the power of the processor and its memory. Finally, the MAC sends the processed data to the higher layers, like TCP/IP for additional processing.

The receiver circuits are typically similar or more complex than those of the transmitter. So, if one were to exclude the effect of the PA in the transmitter, then the receiver may in many cases consume more power than the transmitter. However, of course, in almost all cases there is a PA in the transmitter, so the transmit power is almost always higher than that of the receiver.

3.2.3 Power Consumption in Sleep Mode

Most IEEE 802.11 stations support some form of sleep mode, where many of the transmitter and receiver circuits are in a low-power mode. This can come in very handy along with some 802.11 protocols for power savings. For example, the station can go into sleep mode and only listen to the beacons sent by the AP, to see if there is data queued up for the AP to send to the station. Also, the protocol supports the station listening every n-th beacon (e.g. every fourth beacon). Between beacons the station can go into sleep mode and save power consumption.

Typically, a station can switch quickly between transmit and receive modes. This is important for 802.11 operation. However, it can take some time to enter and exit sleep mode, since there is often a time delay in putting circuits into sleep mode and bringing them out of sleep modes.

Often devices support several sleep modes. Here we will use the terms sleep mode and deep sleep mode. The sleep mode consumes more power than the deep sleep mode since more circuits are powered on. However, it takes less time to enter or exit sleep mode than it does to enter or exit deep sleep mode. This is because in sleep mode some of the circuits remain powered on, while those same circuits of powered down in deep sleep mode.

3.2.4 Power Consumption in Deep Sleep Mode

Some IEEE 802.11 stations support a very low-power sleep mode, which we will call "deep sleep." Other names may be used in different implementations.

To get to the lowest power deep sleep mode almost all circuits need to be turned off. One of the main circuits that may be still on in sleep mode but is turned off in deep sleep mode is the processor memory. The memory may contain a lot of software code and after being turned on, it may take a while to load that memory from non-volatile memory, like Flash memory. This can take over 10 ms and sometimes longer, depending on the implementation. However, by turning the memory off, versus putting it in some form of standby mode, significant power consumption savings is possible.

Other circuits like the accurate clock circuit can be turned off and replaced with a lower power, but less accurate, clock circuit. This can save power, but it decreases the accuracy of the time when receiver transitions from deep sleep to receiver mode, to listen for a beacon. To compensate the station typically transitions from deep sleep to receiver mode early to accommodate any inaccuracies in the low-power clock. This can cause the station to spend additional time in receive mode, consuming additional power.

Another factor that designers consider when designing for an ultra-low-power deep sleep mode is the leakage current in the circuits. Even with the circuits off, there can still be a leakage in the circuits. The level of leakage depends on the specific semiconductor technology process being used for the IEEE 802.11 implement. Once the vast majority of the circuits are turned off, the remaining power from the leakage current must be considered.

The exact power consumption levels of the sleep and deep sleep modes depend on implementation details, so all we will say here is that the power consumption of the deep sleep mode can be 10 times lower than that of the sleep mode. So, for ultra-low-power consumption devices use of the deep sleep mode can be very beneficial. However, there is the latency (delay) it takes for a station to transition from deep sleep mode to receive mode, which we said can easily be more than 10 ms. So that needs to be taken into consideration when using the deep sleep mode.

Next, we look at how a wake-up receiver can be used to save power consumption at the IEEE 802.11 station.

3.3 Wake-up Radio Concept

There are a number of low-power use cases where the user needs very long battery life and also needs a fast response, so the user does not sense a long delay in operation. One such application is home automation where the user wants to control household items and wants to get a quick response to control messages. For example, if the user wants to turn on or off a device like a light, or the user wants to open or close the blinds on the windows. In almost any application where the user needs a fast response but cannot tolerate a short battery life in the sensor or actuator, a WUR is an excellent way to meet this dual requirement of fast response time and long battery life.

With traditional 802.11 operation one can possibly fulfill one out of two of these requirements, but typically not both. For example, if the 802.11 station never goes into sleep more or deep sleep mode, then the responsiveness of the station can be very fast. However, clearly in that case the battery life could be quite short. The dual of this case is to let the 802.11 station go into deep sleep mode for an extended periods of time and only wake-up from deep sleep every 10 second or so. In this case, the battery lifetime could potentially be quite long. However, clearly with this approach the response time to the user could be up to 10 seconds or more, which could be very undesirable for the user.

The WUR is an excellent solution to this problem of having long battery lifetime and fast user response.

What is a WUR? It consists of several components. At the core, is an ultra-low-power wake-up receiver in the IEEE 802.11 station. The power consumption of this receiver is typically under 1 mW in receive mode and could be even lower, say several 100 µW. This is dramatically lower than the power consumption of a regular 802.11 station. Since we know from above, that the RF circuits can be some of the major power consumers of the receiver mode, an ultra-low-power RF receiver is required. We will not go into specific ultra-low-power wake-up receiver RF architectures here, and we will make a few points about the characteristics of such receivers. Typically, the noise figure of the low noise amplifier (LNA) in a wake-up receiver is higher than that of the LNA in a regular 802.11 receiver. This allows the wake-up receiver to save significant power on the LNA, which is a large consumer of power. The accuracy of the LO can also be lower than that of the regular 802.11 receiver, allowing power savings in the frequency synthesizer. Many ultra-low-power wake-up receivers use a non-coherent receiver which allows for the use of an envelope detector and only a single mixer, BB filter, and DAC. With all these power savings methods the RF power consumption can be significantly reduced. However, the standard for the WUR must account for these power savings methods.

That brings us to what needs to be done in the PHY and MAC layers for the WUR. The PHY layer must accommodate the non-coherent receiver with the proper modulation. Since the noise figure of the LNA is higher than in a regular 802.11 receiver, the PHY must accommodate this difference by lowering the data rate in order to maintain a good operating range. The PHY design decisions to accommodate the ultra-low-power RF wake-up receiver are described in Chapter 4.

Two of the fundamental challenges that the MAC must accommodate for this ultra-low-power RF wake-up receiver are to deal with the lower data rate and the need to provide timing synchronization so that the wake-up receiver itself can be duty cycled, to save power consumption even further. To deal with the lower data rates the MAC needs to design MAC frames that use much fewer octets (bytes) than a typical MAC frame. This is so that the packet does not last too long in time, since not only would that cause the wake-up receiver to be on for a longer period of time, but also the access point would be sending 802.11ba packets for an extended period of time, which would prevent other uses during that time, so the packet duration must be kept reasonable. The other challenge the MAC needs to solve is to provide time synchronization using only 802.11ba MAC frames and protocols, so that the wake-up receiver can be placed into its sleep mode for periods of time and then return to wake-up receive mode to listen for wake-up messages. There are a number of details of the MAC layer design, which are covered in detail in Chapters 6 and 7.

Let us say we have an 802.11 station that has the regular transceiver (transmitter and receiver), for example an 802.11n transceiver as well as an 802.11ba receiver. And the 802.11 access point has the regular transceiver, once again 802.11n, as well as the 802.11ba transmitter. We can now describe the timeline of the WUR operation. There are a number of steps it takes in order to get into this operating mode, which we will not discuss here, since those steps will be discussed in detail in Chapter 6. Here we assume the regular transceiver has been put into deep sleep mode, and the 802.11ba wake-up receiver is listening for a wake-up message from the 802.11ba transmitter in the access point.

In Figure 3.3 we show the regular transceiver in the AP, the wake-up transmitter in the AP, the regular transceiver in the low-power device, and the wake-up receiver in the low-power device. We show the operating modes transmissions and receptions from the various radios in the AP and the low-power device. We do not concern ourselves with the operating modes in the AP since it is not the

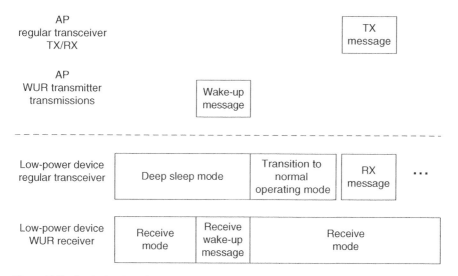

Figure 3.3 Basic timeline for wake-up radio (WUR) operation.

low-power device. We see that the low-power device has been placed in deep sleep mode using very little power. It may operate in that mode for an extended period of time. The 802.11ba wake-up receiver is in receive mode listening for a wake-up message addressed to it. We see that the 802.11ba transmitter in the AP sends a wake-up message. The 802.11ba wake-up receiver processes that message and if the wake-up message is addressed to it, then it sends a signal within the low-power device to wake-up the regular transceiver. We see in the figure that it takes some time for the regular receiver to transition from deep sleep mode to normal operating mode, where it can transmit and receive packets. After that period of time the AP sends a message to the low-power device over the regular transceiver. After that point there may be a sequence of messages and eventually the regular receiver in the low-power device may transition back to deep sleep mode.

The power savings in this example comes from the fact that the power consumption of the WUR receiver is significantly less than that of the regular receiver, so it does not consume much power listening for a wake-up message. More details of the protocols are covered in the MAC chapters.

In the example above the 802.11ba WUR receiver is active all the time listening for a wake-up packet. So, the power consumption is dominated by the active receive power of the WUR receiver, which we said earlier is typically several hundred microwatts.

There is a method of even consuming less power consumption at the low-power device, and that is to duty cycle the wake-up receiver. Let us say, for example, that the wake-up receiver is on for 4 ms for every 200 ms. So, its duty cycle is $\frac{4}{200} = 2\%$, meaning the contribution of the active receive power of the WUR is 2% of the active receiver power, which can be quite low.

Figure 3.4 shows an illustration with a 50% duty cycle for the wake-up receiver. We see that the wake-up packet is sent when 802.11ba wake-up receiver is in receive mode. Both the AP and the STA know the duty cycle operation, so the AP knows to only send a wake-up message to the wake-up receiver when it is in receive mode and not while it is in sleep mode.

In order to duty cycle, the wake-up receiver AP and the STA need to set up this duty cycle using a MAC protocol, which will be described in the MAC chapters. In that way, the AP and the STA have a common knowledge of when the wake-up receiver is in active mode, so the AP can send a wake-up message during the time the wake-up receiver is in active mode. That way the wake-up receiver is able to receive and process the wake-up message.

One more key aspect of this duty cycle mode is for the AP and the STA to maintain synchronization. Both the AP and STA each have their own local clock for maintaining a sense of time. The accuracy of the AP local time synchronization function (TSM) clock is ±100 ppm (parts per million), which means the error in

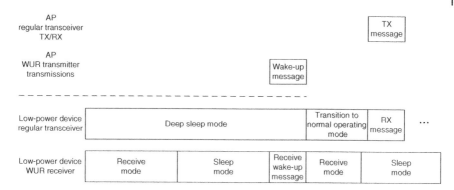

Figure 3.4 Duty cycled timeline for wake-up radio with duty cycle operation.

the clock is off from an ideal clock by at most 100 parts in 1 million. In other words, after 1 second the worst-case timing error is ±100 μs. Here the ± symbol means the clock can be up to 100 μs ahead or up to 100 μs behind an ideal clock, after that 1 second. Similarly, the TSF timer in the 802.11ba STA is also ±100 ppm, so if we consider the worst-case condition where the AP clock is fast and the STA clock is slow, then the clock offset between the AP and the STA is as much as ±200 ppm, meaning that in 1 second the difference between the AP clock and the STA clock can be as much as ±200 μs. This needs to be considered for the duty cycle operation, so that when the AP sends the wake-up message the station is in active mode and can receive the message.

Let us look at what happens if after the AP sets up the duty cycle operation of the 802.11ba STA, it does not send a wake-up message to the STA for 10 minutes. Well that is 600 seconds, and during that time, the clocks between the AP and the STA could drift apart by up to 120 ms, which is much longer than the active time of the 802.11ba wake-up receiver of 4 ms. We clearly do not want the 802.11ba receiver to be in active mode for twice the clock drift time, which would be 240 ms. That would actually be longer than the 200 ms, between the 4 ms active times, in the duty cycle operation. So, we need some method to maintain synchronization between the AP and the STA, so their clocks do not drift too far apart. The solution to this problem is the WUR beacon, which can be sent out on a regular schedule, and from that beacon the 802.11ba STA can resynchronize with the AP clock. This is similar to the standard 802.11 beacon, which is used for synchronization, as well as many other functions.

One thing that is unique about the WUR beacon is it does not need to be sent every time that the 802.11ba STA is in active mode, so in the example above, it does not need to be sent every 200 ms. The reason this was done is that when the AP sends this beacon, other STAs cannot use the medium (air waves). So, the standard was designed to allow the beacon to be sent out less often and use up less medium time. Using the example of having the wake-up receiver be in active mode every 200 ms, the beacon could be sent every five of those periods, or every second. Now, if the beacon is sent less often than the clocks can drift more between beacons, so the active receiver time needs to be long enough to accommodate the worst-case clock drifts, as well as the duration of the 802.11ba packet, so that the wake-up receiver can receive the full packet.

Now, given this duty cycle mode of operation, the overall power consumption of the low-power station can be significantly lower than if the 802.11ba wake-up receiver is in active mode all the time. In Section 3.4 we will give an example of the power consumption using this duty cycle mode.

But before we get to power consumption numbers, we need to talk about latency (delay). The duty cycle mode increases the latency of operation. For example, if duty cycle mode is not enabled, then when there is a message at the AP ready to wake up the low-power device, the AP can send the wake-up message as soon as it has access to the medium, which can be a few milliseconds or less. Then we have the duration of the wake-up packet, which is a millisecond or less, and then the time it takes for the regular 802.11 radio to transition from deep sleep to active mode, which as we mentioned earlier is typically tens of milliseconds. So, we are looking at a latency of typically tens of milliseconds if the wake-up receiver does not use duty cycle mode. This provides very fast response to the user. Say the user is activating a switch of changing the setting of a device in a home automation system. With this type of latency, the action would occur very quickly after the user selected the setting on, say, a smartphone application.

However, if the duty cycle mode is use then the worst-case latency increases by the duration of the duty cycle period, which in our example above is 200 ms. The reason for this is the worst-case condition is that the message reaches the AP just after the wake-up receiver transitioned from active mode to sleep mode. Then the AP needs to wait until the next time the wake-up receiver is in active mode, which is 200 ms in the example we have been using. Now the average increase in latency is only half of that (100 ms) since the time that the message reaches the AP and it needs to send a wake-up message is typically random, and can fall anytime during that 200 ms interval. So, we see duty cycle mode saves power consumption at the cost of some increase in user-perceived latency. One of the great things about the wake-up receiver is that we can use duty cycle periods of only a few 100 ms, and still consume small amounts of power. That is difficult to do in a regular 802.11 STA, since the active power consumption of a regular receiver is much higher, and it is difficult to use deep sleep mode over these short periods of time, when the time to transition in and out of deep sleep is itself tens of milliseconds.

In our example above, if we have a worst-case latency of 200 ms, plus the duration of the WUR packet, plus the time it takes the regular receiver to transition from deep sleep into active mode and receive a message, we are looking at a worst-case user-perceived latency of around 250 ms, and an average user-perceive latency of around 150 ms. This is quite acceptable in many use cases, and even a little bit longer latency may be okay in some use cases.

Next, we will look at an example of power consumption when using a WUR.

3.4 Example of Power Consumption Using a Wake-up Radio

A WUR is useful for battery-operated devices which require low latency (delay). Here we will look at the power consumption of a WUR use case as a function of the duty cycle of the WUR. We will consider the effect of the sleep power consumption as well as the power consumption of the wake-up receiver. What we will not consider here is the power consumption of exchanging data over the main radio. The reason for that is threefold. First, the power consumption for exchanging data over the main radio is heavily dependent on the use case, and there are many possible use cases. Second, for many low-power applications, the frequency of exchanging data over the main radio can be infrequent, and so it may not have a major impact to the overall power consumption. Finally, that part of the power consumption is the same with or without a WUR, so it makes sense to focus here on the effect of the WUR.

To give an example, we need to consider a generic set of example power consumption values, which do not necessarily correspond to any specific implementation. Having gone through one example it is straightforward to change some of the power consumption values and rerun the evaluation with new numbers. Here we assume that the main radio has a deep sleep mode, which is its lowest power consumption mode. This is quite typical for an implementation. As mentioned earlier, there is typically a delay from transitioning from deep sleep into active mode. This is added to the latency of the response to the wake-up message. So, to meet an actual latency response target, the WUR latency may need to be lower than the application-layer latency requirement so that the overall latency meets the application-layer latency requirement. If the main radio does not have a deep sleep mode and only has a regular sleep mode, it is straightforward to replace the power consumption of the deep sleep mode with the power consumption of the regular sleep mode. Of course, that will typically lead to higher overall power consumption values.

For this analysis we only need three parameters. First, we need to the power consumption of the deep sleep mode of the main radio. Next, we need the active power consumption of the wake-up receiver, which is the power consumption when the WUR is in receive mode. Finally, we need the duty cycle of the WUR. This duty cycle is typically based on the worst-case latency that is acceptable to the user. If the acceptable latency is very low, then the wake-up receiver needs to be on frequently to support this low latency requirement. If, on the other hand if a higher application-layer latency is acceptable then a lower duty cycle value can be used, leading to a lower overall power consumption. Here we also assume that when the WUR is in sleep mode it does not add significant power to the deep sleep power consumption. This is typical since deep sleep covers the leakage current in the chip and the power consumption of a low-power clock.

For our example we will use some generic values for a low-power internet-of-things (IoT) Wi-Fi device. These do not apply to any specific implementation. We will say that the deep sleep power consumption is 10 µW. Since the 802.11 project authorization request (PAR) stated that the wake-up receiver power consumption is less than 1 mW and we consider this an upper bound on the 802.11ba receiver active power consumption, we will use half of that in our example. So, the active receive power consumption for the WUR will be 500 µW in our example. These example values are summarized in Table 3.1.

Here we have chosen to give this example using power measured in µW. Some devices specify power in µA (or mA), which is the current drawn from the battery. This same process works just as well for that approach of measuring power consumption. The final parameter we need is the duty cycle which we will indicate simply by DC, which is a value between zero and one. The duty cycle measures the fraction of the time that the wake-up receiver is in active mode. So, for example, if the wake-up receiver was in active mode for 5 ms every 100 ms, then the duty cycle would simply be 0.05, which we will also refer to as a 5% duty cycle.

The average power consumption is simply the deep sleep power consumption (which is on all the time) plus the wake-up receiver power consumption times its duty cycle.

$$P = P_{DS} + DC \times P_{WUR}$$

So when the wake-up receiver is on all the time, the power consumption is $P = P_{DS} + P_{WUR}$ and as the duty cycle approaches zero, the power consumption is approximately $P \approx P_{DS}$.

Figure 3.5 shows the average power consumption for the example we provide here, where the wake-up receiver duty cycle varies between 0 and 100%.

Table 3.1 Example power consumption values.

Parameter	Symbol	Example value
Deep sleep power consumption	P_{DS}	10 µW
Wake-up receiver active power consumption	P_{WUR}	500 µW

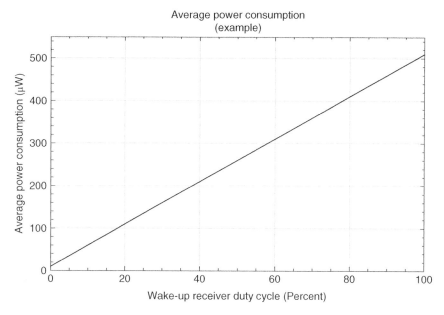

Figure 3.5 Average power consumption for example.

We see that the power consumption does not go below the deep sleep power consumption since that is on all the time. And then the power consumption increases with the increased duty cycle.

From this example, we see that duty cycling the 802.11ba receiver is an excellent way to ensure very low-power consumption.

So, it is important to understand what affects the duty cycle value. We will discuss the choice of duty cycle values in Section 3.5.

3.5   Selection of Duty Cycle Values

Clearly, a lower 802.11ba receiver duty cycle lowers power consumption. When the deep sleep power consumption is very low, like in a low-power Wi-Fi design, duty cycling the 802.11ba receiver can provide significant power consumption savings, as shown in the previous section. However, there are factors that must be considered when selecting the duty cycle value.

When the 802.11ba receiver is duty cycled it alternates between two modes of operation: active receiver mode and sleep mode. See Figure 3.6.

For convenience we have labeled the time when the 802.11ba receiver is in active receive mode as T_{On} and the time that the 802.11ba receiver is in sleep mode

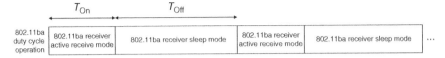

Figure 3.6 Modes of the IEEE 802.11ba receiver duty cycle.

as T_{Off}. It is important to look at each of these periods separately since there are factors that affect the choice of each of these values.

The 802.11ba receiver duty cycle is simply,

$$DC = \frac{T_{\text{On}}}{T_{\text{On}} + T_{\text{Off}}}$$

There are several factors that affect the choice of the T_{On} duration. First, the receiver must be in active receive mode longer than the 802.11ba PPDU duration; otherwise, it could not receive the PPDU. The value of the PPDU duration depends on the data rate, either the low data rate (LDR) or the high data rate (HDR), as well as the size of the MAC frame being carried in the PPDU. A typical MAC frame in 802.11ba is six bytes. With that frame size, the PPDU duration using the HDR is around one-fourth of a millisecond, and when using the LDR, the duration is around 1 ms.

The second factor that affects the choice of the duration of the active receive time is the drift between the clocks on the AP and STA. As was mentioned earlier both the AP and the STA have a clock accuracy of ± 100 ppm, meaning in the worst-case condition where one of those clocks runs fast and the other clock runs slow, the total clock accuracy is ± 200 ppm. This means that after 1 second the difference in the clocks could drift as much as ± 200 μs and within 10 second the difference in the clocks could drift as much as ± 2 ms. So, we see that the second thing that affects the choice of the active receive duration is how frequently the AP and STA are resynchronized. This resynchronization is performed using a WUR beacon, in which the AP sends out a shortened version of its clock value, and the STA aligns it clock value to match the AP clock value. Details of the WUR beacon are provided in the subsequent MAC chapters.

In the case where the clocks are resynchronized every 10 second, the STA needs to activate its receiver 2 ms earlier than normal and keep it on 2 ms longer. So, if the packet duration is 1 ms, then the duration of the receiver active time is now a total of 5 ms. Clearly, a longer period between WUR beacons increase the required active receive duration, so we do not want a long clock drift between WUR beacons. However, each WUR beacon does use the medium (air waves) so sending it out too frequently could consume too much of the medium time. The regular 802.11 beacon is typically transmitted every 100 ms, which may be too often for the WUR beacon since the clock drift over 100 ms is only ± 20 μs.

So, the AP could be configured for a reasonable WUR beacon interval of around 1 second, which only leads to a clock drift of ± 20 µs, which only increases the time that the 802.11ba receiver is in the active receive mode, a small about. Finally, since 802.11 uses a carrier sense protocol, the AP needs to wait until the medium is available before sending a wake-up packet, or a WUR beacon for that matter, so the channel access time should also be considering when specifying how long the 802.11ba receiver is on. Typically, the channel access time is small compared to the packet duration and the clock drift, so it would not add much to the on time. The exact choice of the WUR beacon interval can be controlled by the 802.11ba MAC layer.

The second duration we need to consider is the 802.11ba sleep duration (or off time). Clearly, the longer the off duration the lower the receiver duty cycle and hence the lower the power consumption. However, there is another factor that affects the selection of the 802.11ba receiver off time: latency. Typically, an application layer affects when the STA is to be woken up. This could be driven by when the user takes an action. This could be in a home automation use case and the user changes a setting on a home-automated device, like automated window blinds. There are, of course, many other use cases. The main point here is that when the user changes a setting, this means that the AP will need to send a wake-up message to the STA so that the appropriate action can be taken. And there is no time synchronization between the user changing the setting and the time intervals when the 802.11ba receiver is in active receive mode (on) and sleep mode (off). So, we can think of the time when the message to wake up the STA being ready at the AP to be any random time during those intervals.

If we are lucky and the message is ready to send just before the 802.11ba receiver enters active receive mode, then the latency, or delay, from the time the message is ready to be sent until it can be sent, is very small. However, if we are unlucky and the wake-up message is ready to be sent just as the 802.11ba receiver goes into sleep mode, then the AP needs to wait for the entire sleep mode duration until the 802.11ba receiver is on before it can send the wake-up packet. And, of course, the message could arrive at another time when we are not too lucky or too unlucky. So we see that the latency, or delay, from when the AP has a wake-up message to send and when it can actually send the wake-up message is really a random variable, which can vary from as low as zero and as high as the duration of the sleep interval. This is a uniform random variable between those two values. In most use cases, we are concerned about the worst-case latency since that affects the longest period of time the user needs to wait between selecting a setting and the action taking place. For some use cases, this needs to be small so that the user does not perceive a long delay. In other use case, a longer delay may be acceptable.

Here we discuss the latency from the time the AP is ready so send the wake-up message until it can send the message, but there are also typically other delays that need to be considered when determining the overall end-to-end delay. In some use cases the sleep duration could be 100 or 200 ms, which may be acceptable to the user, while in other use cases a sleep duration of a second or more may be acceptable. Clearly, the use cases which can tolerate longer latency can have a lower receiver duty cycle and hence a lower power consumption. One of the main advantages of the 802.11ba system is that the active receiver power consumption of the 802.11ba receiver is so low that, in many cases, it is possible to have both low latency and low-power consumption.

3.6 Conclusions

Here we have provided an overview of the primary sources of power consumption for 802.11 systems in transmit, receive, sleep, and deep sleep modes. Then we introduced the WUR concept in which the main 802.11 system is typically put into deep sleep mode to save the maximum about of power, and the 802.11ba system is used to bring the main 802.11 system out of deep sleep mode, when there is data to be exchanged between the AP and the STA. We showed how the 802.11ba receiver could be always on and ready to receive a wake-up message, and we also showed how the 802.11ba receiver could be duty cycled to save more power. We provided a generic power consumption example where one can see how the duty cycle operation affects the power consumption. We ended by describing a number of factors that affect the duration of the 802.11ba receiver active receive on time and the 802.11ba sleep-off time. These factors should be considered when setting up the time parameters of the duty cycle mode. Details of how to set up the duty cycle mode are described in the subsequent MAC chapters.

4

Physical Layer Description

4.1 Introduction

The physical (PHY) layer specifies the waveform that is transmitted over the wireless medium. The wireless medium is the airwaves. At the transmitter, the PHY layer receives the physical layer service unit (PSDU), which comes from the medium access control (MAC) layer, which is right above the PHY layer in the protocol stack. The output of the PHY layer is called the PHY protocol data unit (PPDU). So, the job of the PHY layer at the transmitter is to convert the PSDU into the PPDU. The input to the PHY layer at the receiver is the received PPDU which the PHY layer processes and converts to the PSDU, which it sends up to the MAC layer within the receiver. The PPDU, which is sent over the wireless medium, is sometimes casually referred to as the PHY "packet."

The standard specifies the details of how the PHY at the transmitter converts the PSDU into the PPDU to be sent over the wireless medium. Typically, the standard does not specify how the receiver processes the received PPDU to produce the PSDU. How this is done at the receiver is considered to be "implementation specific" and outside the scope of the standard. When developing the standard, the engineers often consider possible ways the receiver might be implemented, but a specific implementation is left up to the implementor.

The standard specifies how to construct the PPDU in sufficient detail so that different vendors can design and build transmitters and receivers that can exchange these PPDUs. This ability for different implementations to exchange data is referred to as interoperability, which enables multiple vendors to build products for the marketplace which work well with one another.

In this chapter, we will give a description of the PHY layer, which focuses on the details of the PHY layer at the transmitter. The PHY layer is specified in Clause 30 of the standard [1]. In Chapter 5, we will look at PHY layer performance which

IEEE 802.11ba: Ultra-Low Power Wake-up Radio Standard, First Edition.
Steve Shellhammer, Alfred Asterjadhi, and Yanjun Sun.
© 2023 The Institute of Electrical and Electronics Engineers, Inc.
Published 2023 by John Wiley & Sons, Inc.

considers both the transmitter and receiver. In that chapter, we need to make some assumptions about a typical receiver implementation. Actual implementations can vary from the assumptions we make here, so the actual performance can vary for each implementation.

To describe the PHY layer we will begin by describing the requirements that were considered by the members of the IEEE when developing the standard. This is a very important step since it determines the directions which were considered when making decisions about the details of the standard.

Next, we will give information about some of the government regulations which were taken into consideration when developing the standard. Since the standard is intended to apply worldwide, regulations from several geographical regions were considered when developing the standard.

Next, we look at the link budget for the 802.11ba link between the transmitter and receiver. One of the goals set by the engineers developing the standard is that the 802.11ba link budget met the link budget of a typical 802.11 link, so that 802.11ba could reach the same range as the 802.11 range. A number of assumptions about implementation need to be made in order for this link budget analysis to be performed. We will describe those assumptions and why they were made.

One of the major decisions that was made early on in the development of the standard was the form of modulation to be used. The requirements for a very low-power receiver drove the choice of on-off keying (OOK) to allow for the use of a non-coherent receiver, which supports a very low-power implementation. A non-coherent receiver does not need to track the carrier frequency/phase, which eliminates power-hungry circuits in the receiver. The IEEE engineers agreed on an OOK modulation where the "on" signals are constructed using multiple carriers. The term "multicarrier on-off-keying" (MC-OOK) is used to describe this modulation. There are several reasons for this decision, which we will describe.

After laying out all this framework we will next describe the details of the 802.11ba PPDU. The PPDU is divided into a non-wake-up radio (Non-WUR) portion and a wake-up radio (WUR) portion. The non-WUR portion is designed to be received and decoded by non-WUR stations. These are 802.11 stations whose receiver is listening for 802.11 PPDUs other than the 802.11ba PPDU. The WUR portion is, of course, designed to be received and decoded by an 802.11 station whose receiver is listening for an 802.11ba WUR PPDU. We will describe both these two portions of the 802.11ba PPDU in detail.

The use of MC-OOK modulation is unique to 802.11ba and with its use comes a few twists. One of these twists is the need to "randomize" the underlying waveform for the MC-OOK symbols to avoid spectral lines in the power spectral density (PSD) of the transmitted signal. Without this symbol randomization these spectral lines would limit the allowed transmit power in some regulatory domains, so it was added to avoid those limitations.

One variant of the 802.11ba PPDU is the frequency division multiple access (FDMA) PPDU, which allows support for transmission in channels wider than 20 MHz. Modern 802.11 stations support bandwidths of 40 and 80 MHz, and even wider bandwidths. The FDMA PPDU enables support for 40 and 80 MHz PPDUs which allow for different content in each of the 20 MHz subchannels.

The chapter ends with some conclusions and preparation for the following chapter on PHY layer performance.

4.2 Requirements

There were several requirements for the IEEE 802.11ba standard that affect the PHY layer design. The 802.11ba standards project began with the approval of the 802.11ba project authorization request (PAR) document [2] in December 2016. There are a number of sections to the PAR, with the scope of the project being one of the key sections, which is provided here,

> This amendment defines a physical (PHY) layer specification and defines modifications to the medium access control (MAC) layer specification that enables operation of a wake-up radio (WUR). The Wake-up frames carry only control information. The reception of the Wake-up frame by the WUR can trigger a transition of the primary connectivity radio out of sleep. The WUR is a companion radio to the primary connectivity radio and meets the same range requirement as the primary connectivity radio. The WUR devices coexist with legacy IEEE 802.11 devices in the same band. The WUR has an expected active receiver power consumption of less than one milliwatt.

The 802.11ba amendment to the 802.11 standard specifies a new PHY layer and modifications to the MAC layer. The MAC layer "Wake-up frames" only carry control information. This was included so as to limit the scope of the amendment so that the MAC frames do not carry data. The data is carried by the "primary connectivity radio" which conforms to one of the other 802.11 amendments (e.g. 802.11n, 802.11ac, 802.11ax, etc.). The term "primary connected radio" was not used in the final standard, but the concept is useful here to understand that 802.11ba is used only to carry control frames and not data frames. For simplicity we often refer to the "main radio (MR)" as shorthand for the "primary connected radio." Clearly, one of the primary MAC frames is used to "wake-up" the more power-hungry circuitry out of sleep mode so that it can be used to exchange data.

Now in terms of specific requirements that affected the PHY layer design we have three.

First, the range of the 802.11ba WUR needs to meet the range of the MR. The reason for this is straightforward. If the 802.11ba WUR is used to wake up the MR, then it needs to meet the same range. If that is not the case, then there would be stations that are within range of the access point using the non-WUR radio but cannot be reached with the WUR. So, if one of those stations is put into sleep mode, it cannot be woken up using the 802.11ba WUR. So, it would be stuck in sleep mode. That is clearly undesirable, so meeting the range of the non-WUR is very important. Once the Task Group started working on the details of the specification, it needed to be a bit more specific on which 802.11 PHY the WUR standard needed to meet in terms of range. The Task Group interpreted this requirement to mean that the link budget of the 802.11ba would meet that of the 6 Mb/s orthogonal frequency division multiplexing (OFDM) PHY data rate. This turned out to be a challenge for an 802.11ba system for a variety of reasons, one of which is that the receive has to consume less than 1 mW in active receiver mode, which is significantly less than the typical power consumption of an 802.11 receiver. It also turns out, as will be described later, the bandwidth of the 802.11ba PPDU is less than that of a non-WUR PPDU, and in some regulatory domains, this results in lower allowed transmit power, which also affects the link budget.

Second, the 802.11ba WUR must coexist with legacy 802.11 stations operating in the same frequency band. This drives the design of a portion of the 802.11ba PPDU. Since legacy STAs cannot decode portions of the 802.11ba PPDU, one portion is included, which can be decoded by legacy STAs. This allows the legacy STAs to understand that they have received an 802.11 PPDU and based on information in that portion, the STAs know the duration of the PPDU so as to avoid transmitting during that time period.

Finally, the third of these PHY-related requirements is that it is "expected" that the WUR active receive power consumption is less than 1 mW. Now as described earlier, the details of the receiver implementation are left up to the implementor, so one way of interpreting this requirement is that the 802.11ba standard makes it possible to design a WUR receiver that has an active receive power of less than 1 mW. It does not specify exactly how that implementation is designed. We should explain what is meant by "active receiver power consumption." Wireless systems have many operating modes (e.g. transmit, receive, sleep, deep sleep, etc.). When we talk about "active receive" mode that is the mode when the receiver is powered on and either receiving a signal or attempting to receive a signal. The receiver may be duty cycled between active receive mode and sleep mode (or deep sleep mode), so the average power consumption will depend on many factors, not only the active receive power consumption. Later in this section we will describe how this requirement drove several decisions regarding the 802.11ba PHY layer design.

Beyond the PAR requirements there was a general agreement in the Task Group that since 802.11ba WUR was for carrying control information and not data, that the standard should limit the complexity. One of the agreements which was reached was to limit the PHY to only two data rates, which are simply referred to as the high data rate (HDR) and the low data rate (LDR). By only supporting two data rates it simplifies the signaling needed to indicate the data rate of the PPDU.

The Task Group considered models for the low-power wake-up receiver and agreed on a model and simulation approach [3, 4].

In addition to the requirements which are set internally at the IEEE there are also regulations that must be considered when developing the PHY layer design. Those considerations will be described in Section 4.3.

4.3 Regulations

The PHY layer is designed to operate in either the 2.4 GHz or the 5 GHz frequency bands, so it is important to understand how the regulations in those frequency bands can affect the PHY layer design. These are the most popular frequency bands used by 802.11 and so operating in those bands makes practical sense. For example, by operating in the same frequency band it makes it possible to share antennas between 802.11ba implementation and the implementation of the other 802.11 PHYs in the same device.

The IEEE focused on several set of regulations. The IEEE considered the US Federal Communication Commission (FCC) regulations which apply in the United States and are also aligned with regulations in other countries in North America, such as Canada and Mexico, as well as other regions around the world. The European Telecommunication Standards Institute (ETSI) specifies spectrum rules which are adopted by regulatory organizations in counties in Europe and also in other geographic regions. Finally, the IEEE considered the regulations in China. In summary, the IEEE focused on FCC, ETSI, and China regulations.

In order to make sure that the 802.11ba WUR meets the same range as the 6 Mb/s OFDM PHY we need to compare the link budget for these two PHYs. One of the key inputs to the link budget is the maximum allowed transmit power, which is typically controlled by regulations. It is also possible in some cases where regulations are not very limiting that the limiting factor is the maximum output power of the power amplifier (PA) in the transmitter.

In all regulatory domains there is a maximum allowed transmit power that can be produced by the transmitter. In some regulatory domains there is also a maximum allowed transmit power that can be transmitted in a specified frequency bandwidth. Depending on the signal bandwidth this can also limit the allowed transmit power.

If there is a regulatory limit for a specified bandwidth then the wider the bandwidth of the signal the more power can be transmitted. So wider bandwidth is better from a transmit power perspective. However, wider bandwidth generally means higher power consumption at the receiver. So, the requirement for ultra-low active receive power pushes toward lower bandwidth. So, there is a tradeoff between higher bandwidth allowing higher transmit power and lower bandwidth enabling low-power consumption at the receiver. These competing requirements ultimately led to the selection of a 4-MHz bandwidth, which is less than the 20 MHz used by the 6 Mb/s PHY. As will be described later the PPDU also includes a 20 MHz portion which can be decoded by non-WUR stations. So, the PPDU consists of two parts: a 20 MHz part and a 4-MHz part. The 20 MHz part is decoded by non-WUR stations, and the 4 MHz part is decoded by the WUR. For the WUR link budget the relevant bandwidth is 4 MHz.

Since we are ultimately interested in matching the link budget of the 6 Mb/s PHY much of the analysis done in the Task Group was in relation to that PHY. In order to perform that comparison, the focus was on the difference between the 802.11ba PHY link budget and the 6 Mb/s OFDM PHY. So often the focus is on the difference between those two PHY versus the absolute values of the individual PHYs. From a regulatory perspective the 802.11ba PHY has a 4 MHz bandwidth and the 6 Mb/s OFDM PHY has a 20 MHz bandwidth.

The primary aspect of the regulations that needs to be considered is the maximum allowed transmit power, for both the 802.11ba PHY and the OFDM PHY.

The FCC regulations for operation in the 2.4 GHz frequency band allow a maximum conducted transmit power of 30 dBm (1 W) and up to 6 dB antenna gain. This applies equally to both the 20-MHz and the 4-MHz signal. There is a minimum 6-dB bandwidth of 500 kHz, but that is easily met by both signals. There is no specific power-per-bandwidth regulation, so in summary the allowed transmit power for both signals are the same.

The FCC regulations in the 5 GHz band depends on the sub-band in which the station is operating. There are four sub-bands,

- 5150–5250 MHz
- 5250–5350 MHz
- 5470–5725 MHz
- 5725–5850 MHz

In the first sub-band, the maximum allowed transmit power is 30 dBm and the maximum allowed transmit power in 1 MHz is 17 dBm. So, for the 20-MHz signal the maximum allowed power is 30 dBm which is the minimum of 30 dBm set by the overall limit and 30 dB ($17 + 10 \; \text{Log}(20) = 17 + 13$) power-per-bandwidth limit. However, for the 4-MHz signal the maximum allowed power is 23 dBm, which is set by the power-per-bandwidth limit. So, in this sub-band the allowed transmit power for the 4-MHz signal is 7 dB lower than for the 20-MHz signal.

In the next two sub-bands, the maximum allowed transmit power is 24 dBm and the maximum allowed transmit power in 1 MHz is 11 dBm. For the 20 MHz signal this results in a maximum allowed transmit power of 24 dBm. For the 4 MHz signal this results in a maximum allowed transmit power of 17 dBm. So once again there is a 7 dB disadvantage for the 802.11ba signal compared with the 20 MHz OFDM signal. Finally, in the final sub-band the maximum allowed transmit power is 30 dBm and the power-per-bandwidth limit is high, so the 30 dBm allowed transmit power limit applies to both the 20 MHz signal and the 4 MHz 802.11ba signal. So, in this fourth sub-band there is no difference in the maximum allowed transmit power.

A similar analysis was done for both the ETSI and China regulations. These results were all summarized in [5] and reproduced here in Table 4.1.

We see that the difference between the maximum allowed transmit power for the 20 MHz OFDM signal and the 4 MHz WUR signal varies from 0 to 4 dB and in many cases 7 dB.

This means that in many cases the 802.11ba WUR link is at a 7 dB disadvantage compared to the 6 Mb/s OFDM PHY. The WUR PHY design must in some way make up for this loss in 7 dB transmit power. This is one of the factors that is included in the link budget considerations that are considered in Section 4.4.

Table 4.1 Summary of regulatory transmit power limits for 20 MHz and 4 Hz signals for FCC, ETSI, and China.

Region	Frequency band (GHz)	TX power limit (dBm)	PSD limit (dBm/MHz)	20 MHz TX limit (dBm)	4 MHz WUR TX limit (dBm)	Δ (dBm)
FCC	2.4	30	N/A	30	30	0
	5.15–5.25	30	17	30	23	−7
	5.25–5.35	24	11	24	17	−7
	5.47–5.725	24	11	24	17	−7
	5.725–5.85	30	33	30	30	0
ETSI	2.4	20	10	20	16	−4
	5.15–5.25	23	10	23	16	−7
	5.25–5.35	23	10	23	16	−7
	5.47–5.725	30	17	30	23	−7
	5.725–5.85	N/A				
China	2.4	20	10	20	16	−4
	5.15–5.25	23	10	23	16	−7
	5.25–5.35	23	10	23	16	−7
	5.47–5.725	N/A				
	5.725–5.85	30	13	26	19	−7

There is one more regulatory requirement that needs to be considered. The FCC limits the maximum allowed power in any 3 kHz band to 8 dBm. This is intended to limit spectral lines in the PSD. The symbol randomization is used to remove spectral lines from the PSD of the signal. This will be described in a later section.

4.4 Link Budget Considerations

The topic of meeting the same link budget for the non-WUR PHY's was considered in the Task Group [6, 7].

There are primarily four items that affect how the WUR link budget differs from the non-WUR link budget: different transmit power, different receiver noise figure, different modulation, and forward error correction (FEC).

As described earlier, if the station is operating in a regulatory domain where there is a power-per-bandwidth regulation, then the transmit power of the 802.11ba PHY may be less than that of the OFDM PHY. The difference can be either 0, 4 or 7 dB depending on the regulatory domain. In the worst-case condition, the 802.11ba transmit power can be 7 dB lower than the OFDM transmitter, due to lower bandwidth in regulatory domains that have a power-per-bandwidth regulation.

The second factor that affects the difference between the WUR link budget and the non-WUR link budget is the difference in the receiver noise figure. The receiver noise figure is the difference in the signal-to-noise ratio at the output of the receiver compared to an ideal receiver that is only affected by thermal noise. In a high-quality receiver the receiver noise figure can be made to be only few dB, but in order to do that the circuits in the receiver need to consume a significant amount of power. There are many circuits in a receiver that can contribute to the receiver noise figure: low-noise amplifier (LNA), mixer, local-oscillator (LO), etc. To design a receiver with a low noise figure each of these circuits needs to be designed well, and in most cases, that means that the circuit consumes significant power. The goal for the 802.11ba receiver is for the active power consumption to be less than 1 mW, so each of these circuits must consume significantly less than 1 mW. To meet these low-power consumption targets, the resulting noise figure of the receiver must be increased. The exact difference between the 802.11ba receiver noise figure and the OFDM receiver noise figure depends on the implementation of both receivers, so it can vary from case to case. The Task Group wanted to budget for the typical difference in the receiver noise figures, so based on the engineering experience of various radio frequency (RF) engineers the IEEE decided to budget for a difference of 8 dB [5], meaning that the 802.11ba receiver noise figure is 8 dB higher than the OFDM receiver noise figure. This is reasonable since the active receiver power consumption of the OFDM receiver can be many times higher than for the 802.11ba receiver.

The third aspect in how 802.11ba PHY differs from the OFDM PHY is the modulation that is used. As will be explained later in more detail, the 802.11ba PHY used multicarrier on-off keying (MC-OOK), which is a variation on the standard OOK. While the lowest order modulation used in the OFDM PHY is binary phase shift keying (BPSK). The required SNR for demodulating OOK is higher than for demodulating BPSK, even when both types of modulation are decoded coherently. The 802.11ba PHY is designed to enable use of an ultra-low-power non-coherent OOK demodulator. This was the primary reason for selecting a form of OOK modulation. Many example research papers which were referred to when starting the 802.11ba project discussed very low-power non-coherent receivers, with active power consumption less than 1 mW.

The required SNR for a non-coherent OOK demodulation is higher than that for a coherent BPSK demodulation. This is illustrated in Figure 4.1. At a bit error rate of 10^{-3} the difference between the non-coherent OOK demodulation used in the WUR and the coherent BPSK demodulation used in the non-WUR is around 6.8 dB.

Finally, the last difference between the 802.11ba PHY and the OFDM PHY that affects the link budget is the fact that the OFDM PHY uses FEC and the 802.11ba does not. In particular, at the lowest data rate the non-WUR uses a rate-1/2

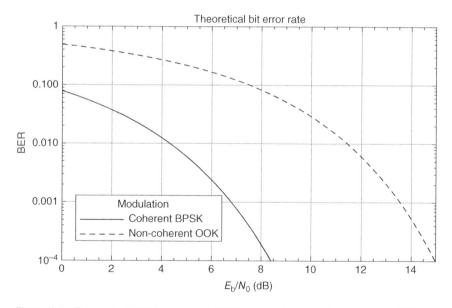

Figure 4.1 Theoretical BER for coherent BPSK demodulation and noncoherent OOK demodulation.

convolution code. There was a proposal to use FEC in 802.11ba [7], but there was insufficient support in the Task Group for this proposal to be adopted.

In summary there are four factors that negatively impact the WUR link budget compared to the non-WUR link budget: possibly lower transmit power, higher receiver noise figure, non-coherent OOK demodulation and no FEC.

Table 4.2 summarizing these differences for the three different maximum allowed transmit power levels. We see that is we consider all the factors the link budget difference is around 27 dB.

The 802.11 PHY design must compensate for all these factors, so the 802.11ba link budget can match that of the OFDM PHY.

The method that is used in the 802.11ba PHY to compensate for all these factors is to use a lower data rate than the OFDM PHY. The 802.11ba PHY supports two data rates: 62.5 and 250 kb/s. The ratio of the lowest rate of the OFDM PHY (6 Mb/s) to the lowest 802.11ba data rate (62.5 kb/s) is 96 : 1. This lower data rate corresponds to approximately 20 dB increase in the bit energy of the WUR compared to the bit energy of the non-WUR. This increase in the symbol energy increases the E_b/N_0 (energy per bit divided by the noise PSD). In the case that the maximum allowed transmit power is the same, this difference in data rate compensates for the other three factors. In cases where the maximum allowed transmit power is either 4 or 7 dB, the implementation may choose to reduce the receiver noise figure to compensate for this transmit power difference.

Table 4.2 Link budget difference between the 802.11ba PHY and the OFDM PHY.

Link budget item	Link budget difference between OFDM PHY and 802.11ba PHY (dB)	Link budget difference between OFDM PHY and 802.11ba PHY (dB)	Link budget difference between OFDM PHY and 802.11ba PHY (dB)
Maximum allowed transmit power	7	4	0
Receiver noise figure	8	8	8
Required SNR for given modulation	6.8	6.8	6.8
Forward error correction	5	5	5
Total link budget difference	**26.8**	**23.8**	**19.8**

4.5 Modulation

One of the key aspects of the PHY layer is that modulation that is used. To meet this goal of ultra-low-power consumption the IEEE decided that the PHY layer should enable the use of a non-coherent receiver which allows the use of a variety of ultra-low-power wake-up receivers which are described in the literature. Many of these receivers work with simple amplitude modulation. In particular, the simplest amplitude modulation is OOK, where the amplitude takes on only two values: On or Off. At baseband, the On symbol the amplitude is set to a nonzero value and the Off symbol the amplitude is zero. Then of course the baseband signal is modulated by a sine wave, so the On symbol become a sine wave and the Off symbol is zero. This simple modulation does not require to receive to know or estimate the exact carrier frequency of phase. It is possible to build a coherent OOK demodulator, so such an implementation is feasible; however, that typically requires higher power consumption.

The IEEE developed a modulation that is a bit more sophisticated than traditional OOK. The IEEE 802.11ba modulation is referred to as MC-OOK where the baseband On symbol is a multicarrier waveform which is constructed as an OFDM symbol, or a portion of an OFDM symbol [8].

Studies of the details of the modulation were considered in the Task Group, including the best bandwidth of the underlying multicarrier waveform [9–14].

There are a number of reasons to choose this modulation. The first reason is rather straightforward. Many of the 802.11 PHY layers are designed using OFDM and so because of this it is quite easy to leverage this OFDM circuitry in the transmitter to construct the MC-OOK On waveform. A second reason is that it is possible to also leverage this OFDM circuitry to construct MC-OOK waveforms operating on different frequencies, making it possible to send multiple MC-OOK waveforms at the same time on different frequencies, which is called frequency division multiplexing (FDM). Since the receivers do not rely on the time orthogonality of these OFDM symbols, we do not refer to this as an OFDM but just the more traditional FDM. The third reason for using MC-OOK versus single carrier OOK, is that with MC-OOK the bandwidth of the waveform can be increased without increasing the symbol rate of the OOK modulation. In single carrier OOK (traditional OOK) the bandwidth of the waveform is proportional to the OOK symbol rate. This is due to the Fourier transform relationship between the time and frequency domain: shorter on-off symbol duration leads to wider bandwidth and visa-versa. However, for MC-OOK we can increase the bandwidth of the waveform by just increasing the bandwidth of the underlying multi-carrier waveform. So, it is possible to increase the bandwidth significantly above the OOK symbol rate. This is advantageous since there are regulatory domains where the allowed transmit power increases with the bandwidth of the signal. So, it is very beneficial to increase the signal bandwidth. If single carrier OOK was used, in order to increase the signal bandwidth, the symbol rate would need to be increased, leading to shorter OOK symbols. These short OOK symbol are then susceptible to intersymbol interference as a result of operation in a multipath channel, which is common in an 802.11 operating environment. Finally, the increased bandwidth provides more frequency diversity, so if there is narrowband fading in the multipath channel, a wider bandwidth signal is less susceptible to that fading. So, for all those reasons MC-OOK was selected for the modulation.

The original OFDM PHY (often referred to as 802.11a) and the next generation OFDM PHY (often referred to as 802.11n) use an OFDM symbol duration of 4 μs. The OFDM symbol is constructed using a 64-point inverse fast Fourier transform (IFFT) where the subcarrier spacing is 312.5 kHz. Some of the frequency domain coefficients, which are the inputs to the 64-point IFFT, are set to zero. Such a subcarrier is often referred to as a null subcarrier since no energy is placed on the subcarrier. The output of the IFFT is a time domain waveform with a duration equal to the inverse of the subcarrier spacing. So, in this case the output waveform of 64-points at 20 MHz sampling rate, has a duration of 3.2 μs. The last 16 samples are prepended to the output of the IFFT. This is referred to either the cyclic prefix (CP) or the guard interval (GI). This results in a 4 μs OFDM symbol. IEEE 802.11ba uses that 4 μs OFDM symbol as a MC-OOK waveform.

When we get into a little more detail on the construction of the PPDU we will see that the PPDU uses both 4 and 2 μs MC-OOK symbols. The 4 μs MC-OOK On symbol can be constructed as was described above. There are several methods of constructing a 2 μs MC-OOK On symbol. One can use a 32-point FFT and apply an 8-sample GI resulting in a 2 μs multicarrier waveform. It turns out it is also possible to use the original 64-point IFFT circuitry and populate the even subcarriers with zero, this results 64 samples at the output of the IFFT, where the first 32 samples are the last 32 samples are the same. So, one can take the first 32 samples, copy the last 8 samples of those 32 sample and prepend them as the GI and you have a 2 μs MC-OOK On symbol.

So, we have described the process for constructing both a 2 and a 4 μs MC-OOK On waveform, we should say something about the bandwidth of the signal, which is controlled by the selection of the frequency domain coefficients. In non-WUR PPDUs the bandwidth of the signal is either 20 MHz or an integer multiple of 20 MHz (e.g. 40 or 80 MHz). As was mentioned earlier there are several factors that affect the choice of the signal bandwidth for the WUR. The first factor is that the power consumption of the receiver increases with the increase in bandwidth. And of course, the baseband sampling rate increases with signal bandwidth. On the other hand, there are regulatory domains (e.g. ETSI and China) where there is a regulation which allows for increased transmit power for higher bandwidth signals, up to an absolute limit. So, a compromise must be made between lower power operation of the receiver and allowed transmit power, where the allowed transmit power affects the range of operation. The IEEE settled on 4 MHz bandwidth which provides a nice tradeoff between power consumption and allowed transmit power. A final factor that is affected by the signal bandwidth is the frequency diversity of the signal. In a multipath channel environment, there is often frequency-dependent fading of the channel. It is possible to have a large fade in signal power over a narrow bandwidth. So, the choice of 4 MHz bandwidth, compared to something like 1 MHz or smaller, provides a diversity in the frequency domain making the signal less susceptible to narrowband fading due to multipath.

If we were using traditional OOK with a symbol duration of 4 μs for example, it would not be possible to have a 4 MHz signal bandwidth, since the signal bandwidth is proportional to the symbol rate, which is only 250 kHz. But with MC-OOK we can construct the underlying OFDM symbol with 4 MHz bandwidth without having to increase the symbol rate. Since the subcarrier spacing is 312.5 kHz, it one populates the subcarrier indicies between -6 and 6 we obtain a signal bandwidth of 4 MHz (Since $13 \times 312.5 = 4062.5$ kHz ≈ 4 MHz). Recall that in OFDM for a 64-point IFFT the index for the subcarrier indicies run from -32 to $+31$. Since it does not impact interoperability (the ability of stations to exchange information) the IEEE did not standardize the exact method of constructing the MC-OOK symbols. The 802.11ba standard specifies the method described above for building the MC-OOK On symbols, and provides several possible sets of frequency domain coefficients that can be used to construct the 2 and 4 μs MC-OOK On symbols. The 2 and 4 μs MC-OOK Off symbols have a value of zero for either 2 or 4 μs.

Now that we have explained the basics of MC-OOK modulation we can now describe the structure of the PPDU.

4.6 Physical Layer Protocol Data Unit (PPDU) Structure

The PPDU can be partitioned into a wideband (20 MHz) portion and a narrowband (4 MHz) portion. We refer to the wideband portion as the non-WUR portion. We refer to the narrowband portion as the WUR portion. The PPDU is shown in Figure 4.2 where each of the fields are labeled. We describe the construction of each of these fields and their purpose.

4.6.1 Non-WUR Portion of PPDU

The non-WUR portion is designed to be decoded by non-WUR stations. The purpose of this non-WUR portion is to indicate to non-WUR stations that this is an 802.11 PPDU and the duration of the PPDU, so that the non-WUR will defer from transmitting during this 802.11ba PPDU. This non-WUR portion is a standard

Figure 4.2 IEEE 802.11ba PHY protocol data unit (PPDU).

OFDM waveform so that it can be decoded by an OFDM receiver. The non-WUR portion consists of the following fields: legacy short training fields (L-STF), legacy long training fields (L-LTF), the legacy signal field (L-SIG), and a two new fields called the BPSK-Mark1 and BPSK-Mark2.

The design of the non-WUR portion of the PPDU required a detailed study of the behavior of non-WUR legacy stations, and went through a several iterations until it was finalized [15–18].

The L-STF consists of 10 short training fields, each of 800 ns duration, so the full L-STF has a duration of 8 μs. The L-LTF consists of two long training fields, for a total duration of 8 μs.

The L-SIG is a single 4 μs OFDM symbol. The L-SIG indicates the modulation and coding scheme (MCS), which is set to MCS0 (6 Mb/s), and the transmit time (TXTIME) is set to match the duration of the WUR PPDU. Non-WUR stations decode the Length field in the L-SIG and defer transmission based on that field. This prevents the non-WUR station from transmitting during the 802.11ba PPDU avoiding a packet collision, which could cause the 802.11ba PPDU to not be received properly by the intended receiver.

Two new 20 MHz OFDM symbols are introduced in 802.11ba: BPSK-Mark1 and BPSK-Mark2. The reason these symbols are introduced is that 802.11 receiving stations rely on the structure of several of the OFDM symbols after the L-SIG to classify which PPDU type is being received. There have been a number of generations of 802.11 operating in the 2.4 and 5 GHz frequency bands. The original OFDM PHY is referred to as non-high throughput (non-HT) and is also referred to as 802.11a, since that is the name of the amendment to 802.11 which introduced this new PHY. That was followed by the high-throughput (HT) PHY, which is specified in the 802.11n amendment. The very high-throughput (VHT) PHY is specified in the 802.11ac amendment. Most recently the high-efficiency (HE) is specified in the 802.11ax amendment. Details of the PHYs up to and including VHT can be found in [19].

In the HT PHY, the L-SIG signals lowest data rate MCS, referred to as MCS0. That MCS uses BPSK modulation. However, to differentiate from non-HT the two symbols after the L-SIG are not BPSK modulated. In the HT PHY, these two symbols after L-SIG are referred to as the HT-SIG, and it uses a modified version of BPSK where the phase modulation is on the quadrature axis and not the in-phase axis. This modulation is referred to a quadrature binary phase shift keying (Q-BPSK).

Since an HT receiving STA can decode both a non-HT PPDU and an HT PPDU, it needs to be able to classify the PPDU based on the first few OFDM symbols. In the HT PPDU MCS0 is signaled in the L-SIG, so if the MCS signaled in the L-SIG is anything other than MCS0 then the STA knows right there that this must be a non-HT PPDU. However, if MCS0 is indicated in the L-SIG then the receiving HT station checks the structure of two symbols after the L-SIG to decide if the PPDU is non-HT (these two symbols use BPSK modulation) or HT (these two symbols use Q-BPSK modulation).

With the introduction of VHT it was necessary to have another signaling technique in the first few OFDM symbols to differentiate between non-HT, HT and VHT. Once again, MCS0 is signaled in the L-SIG but the modulation of the two symbols after L-SIG is different once again. Instead of using BPSK modulation which would be used by non-HT for MCS0 or Q-BPSK which is used in HT, VHT used BSPK in the first symbol after L-SIG and Q-BSPK in the second symbol after L-SIG.

Finally, the HE PHY used a different approach. First, we should observe that in the three previous PHYs the length field is a multiple of three, when MCS0 is signaled. This is because in MCS0 there are three bytes in each OFDM symbol, and since there are an integer number of OFDM symbols the length field must be a multiple of three. Another way of saying this is to say that the length field is a multiple of three is to say that the length field modulo three is zero. In HE, two of the three possible cases for the length field modulo three are used. The two values used are 1 and 2, so as to not use the same value as is used in VHT. These different values are used to signal the type of HE PPDU. For the purposes of 802.11ba we do not need to go into more detail about those HE PPDU types. The main point is that with HE the PHY moved away from differentiation based on BPSK versus Q-BSPK modulation and moved to a method that involved repeating L-SIG. In HE, the first symbol after the L-SIG is called the repeated L-SIG (RL-SIG) and is the same as L-SIG. This is accomplished by using the exact same code bits as those in L-SIG. It also turns out that the pilot tones are the same, so the waveforms for L-SIG and RL-SIG are identical. This concept was leveraged in 802.11ba.

In 802.11ba the value of the Length field, modulo three, is zero. That differentiates it from the HE PHY, which uses values of one and two. To avoid confusion with VHT the two symbols after L-SIG are modulated using BPSK modulation (hence the name BPSK-Mark). The choice of the code bits to use in BPSK-Mark1 and BPSK-Mark2 was selected to be different than those used in L-SIG. The reason for this is that it was expected that PHY designs after HE would possibly want to reuse the RL-SIG detection technique. So, it was decided that the BPSK-Mark1 and BPSK-Mark2 symbols would be based on the code bits in L-SIG. To make the BPSK-Mark1 and BPSK-Mark2 very different than the RL-SIG, the value of the code bits in BPSK-Mark1 and BPSK-Mark2 are the logical complement of the code bits in L-SIG.

The total duration of the non-WUR portion is 28 µs. Adding up the duration of the subfields: L-STF (8 µs), L-LTF (8 µs), L-SIG (4 µs), BPSK-Mark1 (4 µs), BPSK-Mark2 (4 µs) gives a total of 28 µs.

This non-WUR portion is the part that is decoded by non-WUR receivers. These non-WUR STAs will classify the PPDU as a non-HT PPDU and will respect the Length field in the L-SIG. This means that the STA will not transmit during the time interval specified in the Length field. This will cause these STAs to not transmit during the WUR PPDU. This is important to avoid a packet collision.

The next major portion of the WUR PPDU is the WUR portion which is designed to be decoded by a WUR receiver. The WUR receiver does not decode the non-WUR portion since it does not utilize an OFDM receiver but uses instead an ultra-low-power non-coherent receiver. You may recall the bandwidth of the WUR portion is only 4 MHz in bandwidth compared to 20 MHz for the non-WUR portion. This is to enable a lower-power WUR receiver since one of the factors affecting the receiver power consumption is the receiver bandwidth.

4.6.2 Sync Field

The WUR portion consists of two fields: the Synchronization (Sync) field and the Data field. In an FDMA mode we may add padding after the Data field, so there is actually a third field, called the Padding (PAD) field. We will discuss this when we get to FDMA transmission.

The Sync field enables the WUR receiver to perform three functions: WUR PPDU detection, symbol timing recovery, and identification of the data rate used in the Data field. This is a lot for the receiver to do using this one field. Let us begin with the indication of the data rate of the Data field. This is done by having two structures of the Sync field: one for the HDR and one for the LDR. If the data rate in the Data field is the HDR, then the HDR Sync (HDR-Sync) is transmitted. On the other hand, if the data rate in the Data field is the LDR, then the LDR Sync (LDR-Sync) is transmitted. The receiver determines which format of the Sync field was transmitted in order to know the data rate used in the Data field.

The design of the Sync field was a major part of the standards development, beginning with the initial concept, the idea to use the Sync field to indicate the data rate, and concluding with the final Sync field design. This required many technical contributions in the Task Group [20–45] before the design was finalized.

The Sync field uses 4-MHz MC-OOK modulation, based on 2 µs MC-OOK symbols. The reason for this is that we want the duration of the Sync field to be as short as possible, and with the use of 2 µs symbols there are twice as many symbols in a given period of time than if we used 4 µs symbols. This allows for more choices in the sequence of symbols that can be used. One may ask about the impact of intersymbol interference when using shorter symbols, in a multipath channel. The IEEE simulated the use of these 2 μs symbols using a multipath channel model and the intersymbol interference is not an issue

For convenience we will now indicate the MC-OOK symbols with logical bits of ones and zeros. We will indicate an MC-OOK On symbol as a one (1) and an MC-OOK Off symbol as a zero (0). So {1, 0, 1, 1} indicates {On, Off, On, On}.

The duration of the HDR-Sync is 64 μs because it consists of thirty-two 2 μs MC-OOK symbols. Given the notation described above, we can represent the HDR-Sync using 32 bits. The duration of the LDR-Sync is 128 μs since it consists of sixty-four 2 μs MC-OOK symbols. So, the LDR-Sync can be represented using a sequence of 64 bits. The duration of the LDR-Sync is twice the duration of the HDR-Sync since it needs to operate at a lower signal-to-noise ratio (SNR) than the HDR-Sync. The reason for that is that the LDR Data field is designed to operate at a lower SNR than the HDR Data field, so it is necessary for the same to be true of the Sync field.

So now we just need to specify the bit sequence for the HDR-Sync and the LDR-Sync and complete the description of the Sync field.

The bit sequences for these two fields were designed based on a structure that would make it easy for the receiver to distinguish between the HDR-Sync and the LDR-Sync. If this was not the case, then the receiver could easily become confused as to whether it is an HDR-Sync or an LDR-Sync and then might be unable to decode the Data field.

To accomplish this objective the following structure is used for the HDR-Sync and the LDR-Sync.

$$\text{HDR-Sync} = \overline{W}$$

$$\text{LDR-Sync} = [W, W]$$

where W is a sequence of 32 bits. The notation $\overline{W}$ indicates the bitwise logical complement of W. In other words, to construct $\overline{W}$ from W you convert each logical one into a logical zero and each logical zero into a logical one. The notation $[W, W]$ indicates concatenating two copies of W, resulting in a 64-bit sequence.

This structure accomplishes two goals. First, clearly the duration of the LDR-Sync is twice the duration of the HDR-Sync, enabling it to operate at a lower SNR value. Second, by taking the logical complement of W in the construction of the HDR-Sync we obtain a bit sequence that is quite different. More will be said about detection of the HDR-Sync and the LDR-Sync in Chapter 5.

Another factor in the selection of W was the desire to have a limited run of ones or zeros. A run of N ones indicates that there will be a time period of $2N$ μs in the HDR-Sync field when the transmitter is not transmitting any signal. This because the HDR-Sync field uses the complement of W and so those bits indicated Off symbols which means no transmission. Since each bit corresponds to a 2 μs Off signal, leaving to an Off period of $2N$ µs. The reason to avoid a long duration of no transmission during the Sync field is that other STAs listen to the medium (i.e. air waves) and when they hear an extended period of time with no signal on the medium they could classify the medium as Idle meaning that it can be used by other STAs. Now if that non-WUR STA properly detected the non-WUR portion of the PPDU it will avoid transmission during the PPDU since it respects the time indicating in the Length field. However, there may be STAs that did not decode the non-WUR portion of the PPDU, so for those STAs they could improperly identify the medium as Idle. You will notice in the W bit sequence the longest string of ones or zeros is three which corresponds to 6 µs, which was considered acceptable by the IEEE.

The 32-bit sequence of W is given by,

$$W = [1,0,1,0,0,1,0,0,1,0,1,1,1,0,1,1,0,0,0,1,0,1,1,1,0,0,1,1,1,0,0,0]$$

This exact 32-bit sequence was chosen to have good autocorrelation properties to enable the receiver to correlator for this bit sequence and both detect the presence of the Sync field but also the symbol timing. More will be said about detecting the Sync field in the next chapter. A plot of the autocorrelation of the bit sequence W is given in Figure 4.3. First, we see that with an offset of zero bits, all the bits match and so the autocorrelation value is 32. When the offset increases either to the left or the right, the autocorrelation sequence drops below 16, which would be a value one would expect for a total random bit sequence, where half the

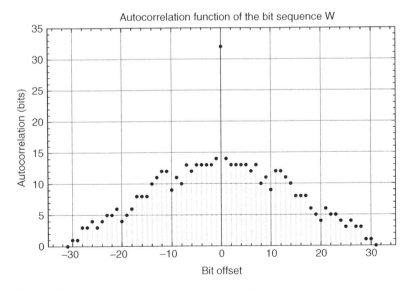

Figure 4.3 Autocorrelation of the bit sequence W.

bits match. Toward the far left and far right the number of bits that match drops off since fewer bits overlap, until we reach zero. The key is that with an offset of zero all the bits match and once we move either to the left or the right less than half the bits match. This autocorrelation property is useful for detecting the HDR-Sync and the LDR-Sync.

As was mentioned earlier, the WUR portion of the PPDU, which is intended to be decoded by the WUR receiver consists of the Sync field and the Data field. Let us next describe the Data field.

4.6.3   Data Field

The Data field carriers the MAC Protocol Data Unit (MPDU) which is the information coming from the MAC layer, above the PHY layer. At the receiver, the PHY layer decodes the Data field and provides the MPDU to the MAC layer at the receiver. As mentioned earlier the Data field supports two data rates: the HDR of 250 kb/s and the LDR of 62.5 kb/s.

Contributions in Task Group related to the design of the Data field covered topics like the number of supported data rates and the use of Manchester encoding [46–49].

The HDR uses a simple Manchester encoding scheme where information bits of ones and zeros are converted to a pair of bits. An information one bit is converted to a pair of bits: {0, 1}. Similarly, an information zero bit is converted to a pair of bits: {1, 0}. There are several benefits to using Manchester encoding. First, it guarantees an equal number of one and zero code bits, which avoids a long string of zero or one code bits. This makes it easy to track the MC-OOK symbols in case there is some timing offset between the transmitter and the receiver. Second, it makes it possible for each information bit to measure the difference between the energy in the first code bit and the energy in the second code bit. This energy difference can be used to decode the information bit. If Manchester coding was not used and the code bit was the same as the information bit, then to decode that bit the receiver would need to compare the energy in that bit to some threshold to determine if this bit is a zero or a one. It turns out that the optimum choice of the threshold for this comparison for OOK depends on the SNR, which would complicate the receiver. So, the use of Manchester encoding simplifies the decoding at the receiver.

Since the data rate of the HDR is 250 kb/s the duration of an information bit must be 4 μs (the inverse of the data rate). Since each information bit is encoded in two Manchester code bits, the duration of each code bit must be 2 μs. So, like in the case of the Sync field, the HDR uses the 2 μs MC-OOK symbols.

The data rate of the LDR is 62.5 kb/s, one-fourth that of the HDR data rate. This data rate is achieved by mapping zero and one information bits each to a sequence of four bits. A zero information bit is mapped to the bits {1, 0, 1, 0}. The one information bit is mapped to the bits {0, 1, 0, 1}. This can also be thought of as repetition coding followed by Manchester encoding. In the case of the LDR 4 μs MC-OOK symbols are used, so the duration of each information bit is a total of 16 μs. Since the inverse of 16 μs is 250 kHz, we have achieved the desired data rate of 250 kb/s.

In some versions of IEEE 802.11, before the Data field, there is a Signal field which indicates a number of parameter for decoding the Data field. One of those parameters typically encoded in the SIG field is the MCS, which indicates the data rate. In the WUR signaling of the data rate is in the Sync field. Another parameter typically indicated in the SIG field is the length of the Data field. However, the WUR PPDU does not have a SIG field, so there is no place to indicate the length of the Data field. In the WUR standard the length of the Data field is known by decoding the first few bytes of the Data field, since the length is encoded in the MPDU. This requires some "cross layer" operation at the receiver since to complete decoding of the PHY layer Data field, it is necessary to decode some of the bits in the MPDU at the MAC layer, to know the length of the Data field. This slight increase in complexity eliminates the need for a SIG field in the WUR PPDU.

4.7 Symbol Randomization

One of the unique aspects of the WUR PHY is the use of MC-OOK symbol randomization. It turns out that if the exact same MC-OOK On symbol was used for each On symbol, the PSD of the resulting waveform would contain spectral lines [50], and a number of studies were performed to eliminate these spectral lines [51–55]. A spectral line, as the name implies, is a very narrow bandwidth in the PSD where there is significantly more power present than at nearby frequencies. Besides the unsightliness of these spectral lines in the PSD these spectral lines have an impact on the maximum allowed transmit power in some regulatory domains. Recall, earlier it was mentioned that the US FCC limits the total power in a 3 kHz bandwidth to 8 dBm. It turns out that the power in one of these spectral lines can be significant, so if they are present the only way to meet the FCC regulation is to lower the overall total transmit power. This would be a major problem, since decreasing the total transmit power would significantly decrease the range of the WUR. The reason these spectral lines occur is that if the same MC-OOK On symbol is used for each On symbol, there is strong correlation between these symbols which are offset in time. This impacts the autocorrelation function of the waveform. Since the PSD is the Fourier transform of the autocorrelation function, these strong correlations lead to the spectral lines.

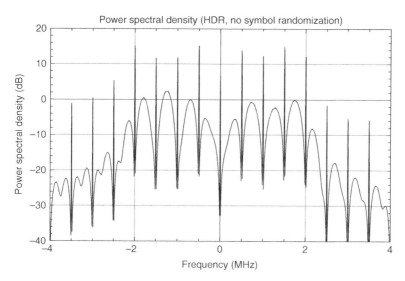

Figure 4.4 Power spectral density of HDR Data field without symbol randomization.

As an illustration of this issue, we can plot the PSD of the HDR and LDR Data fields, without symbol randomization. This is accomplished by simulating a long string of random data bits and them modulating according to either the HDR or LDR modulation scheme. Figure 4.4 shows the PSD of the HDR Data field with random data. There are many spectral lines due to the repeated use of the same 2 μs MC-OOK On symbol. Since the symbol duration is 2 μs, the symbol rate is 500 kHz, so as expected the first spectral line, after DC, is at 500 kHz. Then there are additional spectral lines at multiples of 500 kHz.

Similarly, we can simulate random data and modulate it according to the LDR Data field. Figure 4.5 shows the PSD of the LDR Data field with random data. Here the first spectral line is at 250 kHz, which corresponds to a symbol duration of 4 μs, which is what is used in the LDR Data field.

To eliminate these spectral lines, we need to remove these strong temporal correlations. This is accomplished using MC-OOK symbol randomization. Figure 4.6 shows the symbol randomizer. There is a linear feedback shift register (LFSR) which is initialized with all zeros. For each new symbol, the LFSR is updated shifting the one step through the LFSR. One of those bits is converted from a bit to an integer, m, with a value of either $+1$ or -1. The input waveform is multiplied by this integer value of either $+1$ or -1, which implements a random phase modulation on each symbol. This has been shown to eliminate the spectral lines in the PSD. Secondly, three of the bits (b2, b1, b0) are converted to an integer, n, with a value between zero and seven. This integer is used to look up a cyclic shift value which is applied to the

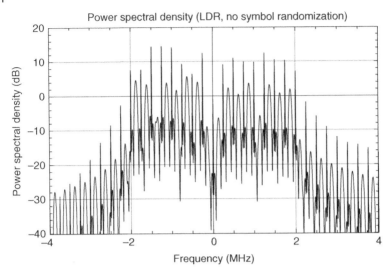

Figure 4.5 Power spectral density of LDR Data field without symbol randomization.

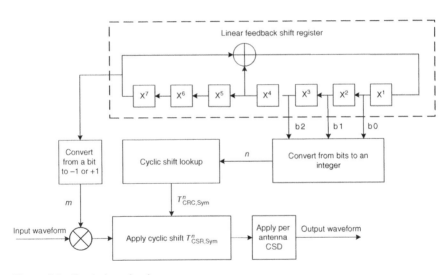

Figure 4.6 Symbol randomizer.

waveform. This has been shown to flatten the spectrum of the PSD. This is useful, since it enables a higher output power in regulatory domains which follow European Telecommunication Standards Institute (ETSI) regulations. In particular, the maximum allowed transmit power in any 1 MHz frequency band, is limited to 10 dBm. So, a flatter PSD leads to higher allowed transmit power in those regulatory domains, which of course leads to longer operating range.

We can run simulation again, this time with the symbol randomizer, used to eliminate spectral lines and to flatten the PSD. Figure 4.7 shows the PSD of random data modulated by the HDR Data field format with symbol randomization. We see the spectral lines have been removed and the spectrum also looks reasonably flat.

Similarly, Figure 4.8 shows the PSD of random data modulated by the LDR Data field format with symbol randomization.

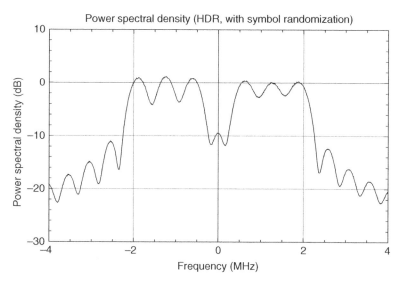

Figure 4.7 Power spectral density of HDR Data field with symbol randomization.

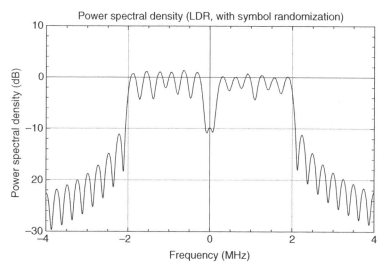

Figure 4.8 Power spectral density of LDR Data field with symbol randomization.

4.8 FDMA Operation

An optional feature of the standard is the ability to send multiple WUR signals in parallel in different 20-MHz channels.

The WUR signal occupies a 20 MHz channel. The non-WUR portion fills the 20 MHz channel, and the WUR portion operates in the middle 4 MHz of the 20 MHz channel.

There are 802.11 access points that support 802.11n and 802.11ac that provide for wider bandwidth channels. In particular, it is quite possible and also quite common to support 40 and 80 MHz channels. In particular, the 5 GHz band has significant bandwidth available and so operating with 40 and 80 MHz channels in that band is quite possible.

The 802.11ba standard supports an optional feature of FDMA where a 40 or 80 MHz channel can be utilized.

4.8.1 40 MHz FDMA

When operating in a 40 MHz channel the wideband portion of the PPDU is duplicated on both 20-MHz channels. The Sync field may either be the same, if the WUR signal in each 20-MHz channel has the same data rate, or the Sync filed may be different, if the WUR signal in each 20-MHz channel has a different data rate. Similarly, the duration of the Data fields in the two channels may be the same, for example if they have the same data rate and the same MPDU duration, or otherwise they may be different. To avoid having the transmission duration different in the two channels, padding is added to the shorter duration signal to match the duration of the longer duration signal. The padding is simply a string of one bits modulated according to the HDR Data field format. Each one bit modulated according to the HDR Data field, corresponds to a 2 μs MC-OOK Off symbol followed by a 2 μs MC-OOK On symbol. This is repeated over and over again to pad the shorter signal to match the duration of the longer signal.

If padding was not used, then the signal in the two channels could have different durations, which can lead to issues. For example, if the transmission on one 20 MHz subchannel abruptly ended while there is still a transmission on the other 20 MHz subchannel, another station whose primary 20 MHz subchannel is the same as the channel where the WUR PPDU transmission abruptly dropped, may then sense that the channel is not occupied and could then transmit on not only the channel whose PPDU transmission just stopped but also the other 20 MHz subchannel, causing a collision with the WUR PPDU transmission corrupting the PPDU preventing the intended receiver from properly decoding the PPDU.

Some STAs with WUR receivers will be operating on one of these 20 MHz channels and other STA with WUR receivers will be operating on other 20 MHz channel. They will process the WUR signal in the channel in which they are operating. This allows different STAs with WUR receivers to receive different WUR signals, and hence different MPDUs from the WUR transmitter, at the same time.

4.8.2   80 MHz FDMA

It is also possible to perform FDMA in 80 MHz channel. Here there is a WUR signal on all four 20-MHz channels. The wideband portion is duplicated in each of these four channels. Similar to the 40-MHz FDMA case, padding is used to ensure that the duration of the signal in each of the channels is the same.

In 80-MHz FDMA we have one other twist. Not all four channels need to be used. It is possible to puncture one of the channels, and not send anything on that channel. So, in essence there are three 20-MHz signal, which are sent on three out of four of the 20-MHz channels within the 80-MHz channel.

4.9   Additional Topics

A few more topics will round out the description of the PHY layer. First, we need to understand that in some of the 5 GHz frequency bands, the bands are shared with radars in some geographic locations. There are regulations in those frequency bands that if the 802.11 network detects the presence of a radar it is to change operation to a different operating channel. So, 802.11 networks operating in those frequency bands have radar detectors built in so to automatically detect the presence of a radar. The frequency bands in which radar detection is required are referred to as dynamic frequency selection (DFS) bands. It was observed by the Task Group that the narrowband 802.11ba waveform can trigger such a radar detector [56, 57]. Detection of a radar based on an 802.11ba waveform is called a false alarm, since the detector triggers based on a signal that is not a radar signal. Because of this phenomenon the 802.11ba standard disallows operation in the DFS frequency bands, to avoid causing false alarms by a radar detector.

As we described earlier an 802.11 preamble begins with the L-STF field, which is used by an 802.11 non-WUR receiver to detect the presence of a PPDU. It was discovered in the Task Group that depending on the choice of the subcarrier values in the underlying OFDM waveform, it is possible for the MC-OOK waveform in an 802.11ba PPDU to trigger a false alarm in the L-STF detection circuit within the non-WUR 802.11 receiver [58–60]. A test was added to the 802.11ba standard that the transmitter must pass to ensure that the 802.11ba MC-OOK waveform does not trigger a false alarm in the non-WUR 802.11ba receiver L-STF detector circuit.

4.10 Conclusions

The WUR PHY layer supports two data rates for the ultra-low-power WUR receiver. The WUR PPDU consists of a non-WUR portion which is designed to be decoded by non-WUR stations so that they will know the duration of the WUR PPDU, so they can defer any transmission until after the end of the WUR PPDU. The WUR portion consists of the Sync field and the Data field. The structure of the Sync field depends on the data rate of the Data field. The WUR receiver uses the Sync field to perform packet detection, determination of the data rate, and symbol timing recovery. The HDR Data field uses 2 µs MC-OOK symbols with Manchester encoding to obtain a data rate of 250 kb/s. The LDR Data field uses 4 µs MC-OOK symbols with a combination of repetition coding and Manchester encoding to obtain a data rate of 62.5 kb/s.

In Chapter 5, we will discuss the performance of the PHY layer for both the HDR and LDR data rates.

References

1 IEEE Std 802.11ba (2021). *IEEE Standard for Local and Metropolitan Area Networks: Wireless LAN Medium Access Control (MAC) and Physical Layer (PHY) Specifications. Amendment: Wake-Up Radio Operation.*

2 IEEE 802.11ba (2016). *Project Authorization Request (PAR)*, Approved December 2016.

3 Wilhelmsson, L., Lopez, M. (2017). Discussion of a Wake-Up Receiver Front-End Model. *IEEE 802.11-17/93r1.*

4 Azizi, S., Shellhammer, S., and Wilhelmsson, L. (2017). IEEE 802.11 TGba Simulation Scenarios and Evaluation Methodology Document. *IEEE 802.11-17/188r10.*

5 Shellhammer, S. and Tian, B. (2017). Regulations and Noise Figure – Impact on SNR. *IEEE 802.11-17/365r0.*

6 Cao, R., Zhang, H., and Hua, M. (2017). WUR Link Budget Analysis. *IEEE 802.11-17/399r1.*

7 Shellhammer, S. and Tian, B. (2017). Data Rates and Coding. *IEEE 802.11-17/670r0.*

8 Azizi, S., Park, M., and Kenney, T. (2017). High Level PHY Design. *IEEE 802.11-17/84r0.*

9 Shellhammer, S. and Tian, B. (2017). WUR Modulation and Coding. *IEEE 802.11-17/366r1.*

10 Azizi, S., Park, M., and Kenney, T. (2017). Studies of PER Performance. *IEEE 802.11-17/367r0.*

11 Jia, J., Jian Yu, R., and Gan, M. (2017). Performance Investigations on Single-Carrier and Multiple-Carrier-Based WUR. *IEEE 802.11-17/373r2.*

12 Suh, J., Aboul-Magd, O., and Au, E. (2017). Waveform Generation for Waveform Coding. *IEEE 802.11-17/376r0.*

13 Park, E., Lim, D., and Chao, J. (2017). OOK Signal Bandwidth for WUR. *IEEE 802.11-17/655r4.*

14 Park, E., Lim, D., and Chun, J. (2017). Signal Bandwidth and Sequence for OOK Signal Generation. *IEEE 802.11-17/964r4.*

15 Tian, B., Sun, Y., and Vermani, S. (2017). WUR Coexistence and Packet Format. *IEEE 802.11-17/675r0.*

16 Cao, R., Zhang, H., and Hua, M. (2017). WUR Legacy Preamble Design. *IEEE 802.11-17/647r4.*

17 Azizi, S., Kenney, T., and Kristem, V. (2018). Further Discussion on Content of BPSK Mark. *IEEE 802.11-18/1564r0.*

18 Shellhammer, S. (2020). Comment Resolutions on BPSK Mark Symbols. *IEEE 802.11-20/77r0.*

19 Perahia, E. and Stacey, R. (2013). *Next Generation Wireless LANs: 802.11n and 802.11 ac*, 2ee. Cambridge University Press.

20 Liu, J., Wu, T., and Pare, T. (2017). Preamble Design for WUR WLAN. *IEEE 802.11-17/679r2.*

21 Lim, D., Park, E., and Chun, J. (2017). Signaling Method for Multiple Data Rate. *IEEE 802.11-17/963r0.*

22 Cao, R., Zhang, H., and Shrinivasa, S. (2017). WUR Preamble SYNC Field Design. *IEEE 802.11-17/983r0.*

23 Shellhammer, S., Tian, B., and Verma, L. (2017). Preamble Design and Simulations. *IEEE 802.11-17/991r0.*

24 Azizi, S., Kenney, T., and Fang, J. (2017). Preamble Options. *IEEE 802.11-17/997r0.*

25 Liu, J., Wu, T., and Pare, T. (2017). Follow Up on Preamble Design for WUR. *IEEE 802.11-17/1020r0.*

26 Lim, D., Park, E., and Chun, J. (2017). Follow-Up on Signaling Method for Data Rates. *IEEE 802.11-17/1326r0.*

27 Liu, J. and Pare, T. (2017). Packet Format of 11ba. *IEEE 802.11-17/1340r0.*

28 Cao, R. and Zhang, H. (2017). WUR Preamble SYNC Field Design. *IEEE 802.11-17/1343r0.*

29 Zhang, H., Cao, R., and Chu, L. (2017). 11ba PHY Frame Format Proposal. *IEEE 802.11-17/1345r5.*

30 Chun, J., Park, E., and Lim, D. (2017). Consideration on WUR Sync Preamble. *IEEE 802.11-17/1352r0.*

31 Shellhammer, S., Tian, B., and Verma, L. (2017). WUR Preamble Bit Duration. *IEEE 802.11-17/1354r0.*

32 Shellhammer, S., Tian, B., and Verma, L. (2017). WUR Preamble Evaluation. *IEEE 802.11-17/1355r1*.

33 Park, E., Lim, D., Chun, J., and Choi, J. (2017). WUR SYNC Preamble Design. *IEEE 802.11-17/1611r2*.

34 Azizi, S., Kristem, V., and Kenney, T. (2017). Discussion on Preamble Sequences for Indication of the WUR Rates. *IEEE 802.11-17/1614r1*.

35 Shellhammer, S., Tian, B., and Verma, L. (2017). Dual Sync Designs. *IEEE 802.11-17/1617r1*.

36 Cao, R., Srinivasa, S., and Zhang, H. (2017). WUR Dual SYNC Design and Performance. *IEEE 802.11-17/1618r0*.

37 Sundman, D., Wilhelmsson, L., and Lopez, M. (2017). WUR 128 μs Preamble Design. *IEEE 802.11-17/1665r3*.

38 Cao, R., Srinivasa, S., and Zhang, H. (2018). WUR Dual SYNC Design Follow-Up: SYNC Bit Duration. *IEEE 802.11-18/70r0*.

39 Chun, J., Lim, D., and Park, E. (2018). WUR Dual Sync Performance. *IEEE 802.11-18/73r2*.

40 Kristem, V., Azizi, S., and Kenney, T. (2018). WUR Sync Design. *IEEE 802.11-18/96r3*.

41 Kristem, V., Azizi, S., and Kenney, T. (2018). 2 μs OOK Pulse for High Rate. *IEEE 802.11-18/97r0*.

42 Jia, J. and Gan, M. (2018). WUR Preamble Sequence Performance Evaluation. *IEEE 802.11-18/100r1*.

43 Shellhammer, S. and Tian, B. (2018). Sync Field Bit Duration *IEEE 802.11-18/122r0*.

44 Shellhammer, S. and Tian, B. (2018). Options for Sync Field Bit Sequence. *IEEE 802.11-18/123r0*.

45 Shellhammer, S. and Tian, B. (2018). Concerns about Sync Detector False Alarms. *IEEE 802.11-18/1201r0*.

46 Park, E., Lim, D., and Choi, J. (2017). Multiple Data Rates for WUR. *IEEE 802.11-17/654r3*.

47 Park, E., Lim, D., and Chun, J. (2017). Data Rate for Range Requirement in 11ba. *IEEE 802.11-17/965r2*.

48 Park, E., Lim, D., and Chun, J. (2017). Symbol Structure. *IEEE 802.11-17/1347r3*.

49 Sundman, D., Wilhelmsson, L., and Lopez, M. (2017). Discussion of Possible BCCs for WUR. *IEEE 802.11-17/1394r2*.

50 Shellhammer, S., Tian, B., and van Nee, R. (2018). WUR Power Spectral Density. *IEEE 802.11-18/824r1*.

51 Kristem, V., Azizi, S., and Kenney, T. (2018). WUR PSD Studies. *IEEE 802.11-18/1165r1*.

52 Lopez, M., Sundman, D., and Wilhelmsson, L. (2018). Spectral Line Suppression for MC-OOK. *IEEE 802.11-18/1179r1*.

53 Shellhammer, S. and Tian, B. (2018). Comparison of Symbol Randomization Techniques. *IEEE 802.11-18/1200r0*.

54 Lopez, M., Sudman, D., and Wilhelmsson, L. (2018). Tuning Symbol Randomization. *IEEE 802.11-18/1529r0*.

55 Shellhammer, S. and Tian, B. (2018). Symbol Randomization Simulations. *IEEE 802.11-18/1558r0*.

56 van Zelst, A., Kim, Y., and Tian, B. (2017). False Radar Pulse Detection on WUR Signal. *IEEE 802.11-17/377r0*.

57 van Zelst, A., Tian, B., and Shellhammer, S. (2017). False Radar Pulse Detection on WUR Signals in DFS Channel. *IEEE 802.11-17/1666r2*.

58 Shellhammer, S., Awater, G., and van Nee, R. (2019). False L-STF Detection Issue. *IEEE 802.11-19/1120r0*.

59 Lopez, M. and Wilhelmsson, L. (2019). Study of False L-STF Detections Triggered by MC-OOK. *IEEE 802.11-19/1178r0*.

60 Kristem, V. and Park, M. (2019). Studies on False Detection of WUR PPDU as L-STF. *IEEE 802.11-19/1838r0*.

5

Physical Layer Performance

5.1 Introduction

In this chapter we discuss the performance of the physical (PHY) layer based on a generic non-coherent receiver. The actual receiver in a station is based on the design of the implementer and can vary from one implementation to the other. The IEEE 802.11ba standard does not specify the use of a specific receiver design, but it does place some requirements on the receiver performance.

Here we will begin by giving an overview of the generic non-coherent receiver that will be used to investigate the performance of the PHY layer. The generic receiver is modeled using a software simulation. We will give a description of some of the aspects of the simulation, for example, the modeling of the wireless channel. Then we will provide some simulation results for the 802.11ba PHY layer under various channel conditions.

Since the wake-up radio (WUR) receiver is intended to decode the narrowband WUR portion of the 802.11ba PHY protocol data unit (PPDU), we will be focusing on that part of the PPDU. The wideband Non-WUR portion is intended to be decoded by an orthogonal frequency division multiplexing (OFDM) receiver and so is out of scope of this chapter.

5.2 Generic Non-coherent Receiver

A non-coherent receive does not rely on knowledge of the phase of the received signal for demodulation. A coherent receiver, which relies on knowledge of the phase of the signal, may use a matched filter to optimally demodulate receive signals. However, to have an accurate estimate of phase of the signal requires

IEEE 802.11ba: Ultra-Low Power Wake-up Radio Standard, First Edition.
Steve Shellhammer, Alfred Asterjadhi, and Yanjun Sun.
© 2023 The Institute of Electrical and Electronics Engineers, Inc.
Published 2023 by John Wiley & Sons, Inc.

significant power consumption. The local oscillator at the receiver must have reasonably low phase noise. The receiver must estimate the phase of the signal using some training fields, etc. The IEEE engineers designed the 802.11ba PHY layer to allow use of a non-coherent receiver which is expected to consume less than 1 mW in active receiver mode. This non-coherent receiver operates on the envelope of the receive signal. There are multiple methods of implementing an envelope detector in a receiver. Here we will use a model of an envelope detector to enable simple implementation in a simulation.

Since a non-coherent receiver is assumed in the development of the PHY layer, the choice of amplitude modulation was a reasonable choice. The specific choice of MC-OOK has a number of other key advantages as described in the previous chapter, but the basic idea of amplitude modulation was chosen since there are a number of ultra-low-power wake-up receivers described in the literature. As described earlier, one of the main challenges for 802.11ba is to meet the range requirements, so it is important for the WUR to work at low signal-to-noise ratio (SNR). Given that constraint we want to use binary amplitude modulation, which is why a form of on-off keying (OOK) was selected.

In our generic non-coherent receiver, we will model the receiver at baseband. So, it is assumed that the radio frequency (RF) signal has been converted from RF to baseband using a downconverter. In actual implementations there are a number of ways this can be done. Here we will begin with the baseband signal, as described in the standard. This baseband signal is, of course, a complex signal, which can be obtained from a zero intermediate frequency (zero-IF) receiver or from another similar RF receiver. The generic non-coherent receiver consists of a simple model of a low-pass filter followed by an envelope detector. In the IEEE, there was also a down sampler since it was assumed that the receiver could operate at a sampling rate close to 4 MHz, and the original transmit signal is typically generated at 20 MHz sampling rate. Of course, the actual receiver can operate at a different sampling rate. The IEEE engineers typically simulated the modem (e.g. Sync detector and Data field decoder) at a 4 MHz sampling rate, since a lower sampling rate enables a lower power receiver. The Task Group agreed on a model for the 802.11ba receiver to be used in the evaluation of the PHY layer performance [1].

So, there are three blocks in this generic non-coherent receiver: a low-pass filter, an envelope detector, and a down sampler.

Here we will describe the low-pass filter that was used in our simulations. This was typical of what was used in simulations from other IEEE engineers. The complex baseband signal spectrum spans from approximately −2 MHz to +2 MHz. This complex baseband signal consists of the inphase (real) signal and the quadrature (imaginary) signal. Each of these components (the inphase and quadrature) can be represented as real signals and have a baseband bandwidth of around 2 MHz.

Figure 5.1 Non-coherent receiver model.

So, filter bandwidth for each of these components should be around 2 MHz to capture the signal and filter out the majority of the noise. Since it is expected that this receiver will be very low-power we assume a rather low complexity filter. So, the choice of a third-order Butterworth filter with a 3-dB cutoff of 2.5 MHz was selected for this model. It, of course, is possible to use a higher order filter at the cost of an increase in power consumption.

The envelop detector is used to extract the envelope of the signal, which is the low-frequency signal riding on top of the underlying RF carrier. This is a nonlinear operation that can be implemented in an actual receiver in a number of ways. The classical example uses a diode circuit that only passes current in one direction. In modern RF integrated circuits, there are a number of nonlinear devices that can be used to implement an envelope detector. In our generic model of the envelop detector, we use a simple mathematical model. There are two obvious choices: $abs(x)$ and $abs(x)^2$. It turns out that both choices give similar results in a simulation. We will use $abs(x)$ in our generic model.

Finally, the model down samples from the 20 MHz sampling rate to a 4 MHz sampling rate, through a simple down sample by 5 block. Typically, in the simulations the transmitted waveform is sampled at 20 MHz to fit with the IEEE channel models, so down-sampling after the channel and before the receiver is required if the receiver is modeled at 4 MHz sampling rate.

So, after the envelope detector there is a real signal sampled at 4 MHz.

The non-coherent receiver model is shown in Figure 5.1.

5.3 Simulation Description

The simulator consists of three high-level blocks: the transmitter, the channel, and the receiver. In the simulator the transmitter constructs the transmit waveform. If there are multiple transmit antennas, then it constructs the transmit waveforms for all the transmit antennas. The IEEE modeled multiple transmit antennas since typically the transmitter is within an access point (AP), and it is common for APs to support multiple antennas. On the other hand, since the receiver is intended to be in an ultra-low-power device the simulations assume a single receive antenna. The channel block implements the model of the wireless channel. There are two channel models that are considered: additive white

Gaussian noise (AWGN) channel model and a multipath channel model referred to as Channel Model D. The receiver consists of two parts: the non-coherent RF receiver model described above and the modem, which will be described below. Finally, performance metrics must be extracted from the simulation. We will describe several performance metrics below.

5.3.1   Transmitter Model

The transmitter generates the narrowband WUR portion of the WUR waveform. This includes the generation of the Sync field and the Data field. Simulations are run for both the high data rate (HDR) and the low data rate (LDR). Based on the data rate, either an HDR or LDR Sync field is generated.

The Sync field is represented by a sequence of bits, where the sequence depends on the data rate. The logical one indicates that an MC-OOK On symbol is transmitted and a logical zero indicates that an MC-OOK Off symbol is generated. The duration of the Sync bits is 2 μs, so the transmitter needs to generate 2 μs MC-OOK On and Off waveforms. The method for generating the MC-OOK symbol waveforms is described below.

The Data field is constructed using random information bits. Typically, the number of bits is set to 48, since the fixed length MAC frame is 6 bytes (48 = 6 × 8), though it is also possible to simulated with more bits representing a longer MAC frame. If the simulation is for the HDR, then Manchester encoding is applied to the information bits resulting in 96 encode bits. Then these bits are converted to 2 μs MC-OOK On and Off symbols, using the same 2 μs MC-OOK On symbol as used in the Sync field. If the simulation is for the LDR, then repetition coding followed by Manchester encoding is applied to the information bits resulting in 192 encoded bits. These bits are converted to 4 μs MC-OOK On and Off symbols.

The Sync field and the Data field are concatenated. These samples are typically prepended and appended by padding samples for the simulation.

5.3.2   MC-OOK Symbol Waveform Generation

The generation of the MC-OOK symbols is described here. There are multiple aspects of the generation of these waveforms. First, is the length of the symbol: either 2 or 4 μs. The 2 μs symbols are used in the Sync field and the HDR Data field. The 4 μs symbols are used in the LDR Data field. The next aspect of the construction of these MC-OOK symbols is the symbol randomization which is used to eliminate any spectral lines in the power spectral density. Finally, there is the cyclic shift diversity (CSD) which is applied when there are multiple transmit antennas, to minimize destructive interference at the receiver. We will take up each of these aspects of the construction of the symbol waveforms.

First, let us start with the basic procedure for constructing the MC-OOK symbols based on the OFDM symbol structure. For the 2 μs symbols we begin with a set of subcarrier coefficients, called sequences in the standard. The IEEE standard provides example sequences for the 2 μs symbols in Annex AC. This generates 64 samples, where the first 32 samples and the second 32 samples are the same. This is because in the frequency domain, the odd subcarrier coefficients are zero. We take the first (or equivalently the second) 32 samples. Then if multiple transmitters are provided, the cyclic shift is applied depending on the antenna number. The IEEE does not require specific CSD values, but recommendations are provided in an Annex. Then the last 8 samples of the 32 samples are prepended to the 32 samples giving 40 samples. These samples are referred to either as the cyclic prefix (CP) or as the guard interval (GI).

Finally, symbol randomization is applied, as described in the previous chapter. A 2 μs MC-OOK Off symbol is 2 μs of zeros.

The construction of the 4 μs MC-OOK On symbol is similar to that of the 2 μs symbol but slightly different. For the 4 μs MC-OOK symbol another sequence of subcarrier coefficients is used. The IEEE standard provides example sequences for the 4 μs symbols in Annex AC. The output of the IFFT provides 64 samples. Based on the appropriate cyclic shift, the samples are shifted accordingly. Then the final 16 samples are prepended giving 80 samples. Finally, the symbol randomization is applied.

A 4 μs MC-OOK Off symbol is 4 μs of zeros.

5.3.3 Channel Model

The transmitter provides one or more transmit waveforms which are then processed by the channel model.

The simplest channel model is the AWGN model. The AWGN channel model does not include any multipath channels, so the main aspect is to add random complex Gaussian noise to the receive signal. The level of noise to add depends on the SNR of this simulation. In the IEEE, the SNR definition that was used was the signal power divide by the noise power, and both were sampled at 20 MHz. The sampling rate matters since a higher sampling rate increase the bandwidth, which will affect the bandwidth of the noise power spectral density. The 20 MHz definition for SNR was chosen since that is the definition used for the 20 MHz 802.11 OFDM PHY, and there was interest in comparing the link budget between the 20 MHz OFDM PHY and the new WUR PHY. Since the bandwidth of the actual WUR is 4 MHz, that is another choice that could be considered, but since there was a goal of comparing to the 20 MHz OFDM PHY, the 20 MHz definition of SNR was used. Finally, a random time offset is added so as to test the Sync field detector. When multiple transmit antennas are modeled in the simulation, there are two other aspects that must be implemented. First, a random phase is applied to each transmit waveform from each antenna. This is because at RF there is going to be a difference in the time duration of the path from each transmit antenna to the receive antenna. At the RF level, this time difference corresponds to a phase rotation at the RF carrier, leading to a phase rotation at the baseband signal. Next, before adding the noise to each of the waveforms from each antenna the signal from each antenna is scaled by the square root of the number of antennas: scale by $1/\sqrt{M}$ where M is the number of transmit antennas. The reason for this is we want to have the correct SNR, and if we scale each waveform by the inverse of the square root of the M that is equivalent to scaling the power of each transmission by the inverse of M, giving the same total transmit power. Then independent AWGN is added to each transmitted waveform. Finally, all the M waveforms are added together representing the combining of the signals at the single receive antenna. The reason they are just added together is that there is no multipath channel. Also, a timing offset is added since the receiver must recover the timing of the packet relying on the Sync field.

The other channel that is typically modeled is referred to as Channel Model D. This is a multiple input multiple output (MIMO) channel model developed during the development of the 802.11n standard, which was the first 802.11 standard that supported MIMO. Here in 802.11ba we have multiple input (multiple TX antennas) single output (single RX antenna) channel, typically referred to as a MISO channel. This is just a subset of the MIMO channel, so use of this channel is very reasonable. A detailed description of the channel is provided in [2]. For any given channel between a single transmit antenna, and the receive antenna is a multipath channel, which is modeled by a set of multipath taps (like an FIR filter) where each tap is a complex Gaussian random variable. The channel model provides the statistics of these multipath values. For each simulation, a different channel realization is generated, representing a different channel condition. For each channel, a multipath model affects several aspect of the waveform. First, the total channel gain varies, so some channel realizations have higher attenuation than others. Secondly, there are multiple versions of the transmit signal delayed by the multipath delay and scaled by the multipath gain. This superposition of these delayed and scaled versions of the signal can result in intersymbol interference. Since 802.11ba uses MC-OOK the duration of the MC-OOK symbols is either 2 or 4 μs, so the effect of intersymbol interference is not very significant. If that were not the case, the intersymbol interference would be more significant. In fact, that was one of the reasons for using MC-OOK symbol duration of at least 2 μs. So, it turns out the main effect of Channel Model D is the random variation in the channel gains. Use of multiple transmit antennas addresses this issue somewhat, so it is very beneficial to use multiple transmit antennas whenever possible.

5.3.4   Receiver Model

The receiver model consists of the generic non-coherent receiver model described earlier and the modem. The modem receives the real waveform sampled at 4 MHz and performs a number of functions: detection of the Sync field, determination of the data rate based on the classification of the Sync field, determination of the timing of the beginning of the Data field based on timing recovery from the Sync field, and finally the decoding of the Data field.

The Sync detector detects the Sync field. This involves deciding if there is an HDR Sync, an LDR Sync, or no Sync field in each experiment within the simulation. A simple model of a Sync detector is used in the simulation. In an actual implementation the Sync field detector may be more complex. Here the Sync field detector consists of a correlator and a power normalizer. The HDR Sync field is represented by the sequence of 32 bits, $\overline{W}$, where each bit represents an MC-OOK symbol with a duration of 2 μs. The detector uses a reference signal which is equal to $Y = \left(2\overline{W} - 1\right)$. If the sampling rate is 4 MHz at the receiver, there are eight samples for each bit. For each new sample at the receiver, a vector of 128 samples is captured, representing the sample of a possible HDR Sync field. These samples are multiplied pointwise times the samples of the vector Y. If this is a HDR Sync field, then the output of this summation will be large. The output of this summation is normalized by the sum of the samples of X, to provide an AGC effect. The output of this normalized correlator is compared to a threshold. The threshold is set so that the false alarm rate (FAR) of the detector is sufficiently low when no Sync field present, and there is only AWGN present.

A similar detector is used for the LDR Sync field detector. Since the HDR and LDR Sync fields are both built upon the bit sequence W, it is possible to build a lower complexity Sync field detector by combining some of the circuitry.

The Sync field detector is shown in Figure 5.2. The value of the Y depend on whether this is the HDR or the LDR Sync field detector.

If the Sync field detector detects what it believes is either an HDR or an LDR Sync field, it provides a pointer in the receive signal to the beginning of the Data field and also indicates the data rate: either HDR or LDR. In practice both the HDR and LDR Sync field detectors are running simultaneously each searching for their Sync field. At each instant the receiver must look at the detection statistics from both the HDR and LDR Sync field detectors and decide if the HDR Sync field is present, the LDR Sync field is present, or neither is present.

The Data field decoder decodes each bit, one at a time. For the HDR the sum of the samples of the first 2 μs is compared to the sum of the sample of the next 2 μs. If the sum of the samples of the first 2 μs is larger than the sum of the second 2 μs then this bit is decoded as a logical zero; otherwise, it is decoded as a logical one. Similarly, for the LDR, the Data field decode compares the sum

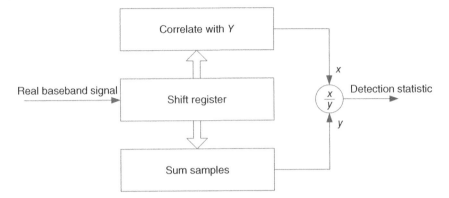

Figure 5.2 Sync field detector used in simulation.

of the samples for the first 8 μs to the sum of the samples for the second 8 μs. If the first sum is larger than the second sum, then the bit is decoded as a zero; otherwise, the bit is decoded as a one. This is performed for all the bits in the Data field. In a real receiver, the Data field decoder must decode the first few bits of the Data field, and using decode those bits into the MAC Length field, to know the duration of the Data field. However, in simulations typically fixed length PPDUs are used.

5.3.5 Performance Metrics

Finally, we need to describe the performance metrics that the simulation needs to provide.

First, we begin with the Sync detector. For each simulation, the Sync detector must decide whether an HDR Sync field is present, an LDR Sync field is present, or no Sync field is present. If the detector decides that either an HDR or an LDR Sync field is present, then it must provide a pointer in the received signal to the beginning of the Data field.

Before simulating with a PPDU the thresholds for the HDR and LDR Sync field detector must be set. That can be done by simulating with 2 ms of AWGN only. The choice of 2 ms was selected in the IEEE to represent a reasonable receiving interval. In practice, the receive window will depend on many factors, like the accuracy of the real-time clock, etc. The thresholds for the HDR and LDR Sync detectors are to be set so that the FAR, based on an AWGN input, is less than 10^{-2}.

Once the Sync detector thresholds are set, there are several performance metrics that can be considered for the Sync detector. There is the Sync detection failure rate for each detector. A detection failure means that the Sync field is not detected when it is actually present. This will depend on the SNR and also the channel model, so typically this is plotted versus SNR for various channel models.

Another Sync detector performance metric is the probability of improperly detecting an HDR Sync as an LDR Sync, or visa-versa. We call this the Sync field classification error rate. This also depends on the SNR and channel model.

Finally, for the Sync detector, we can look at the timing error of the pointer to the beginning of the Data field. This is important since even if the Sync field was detected correctly, if the timing recovery is poor, then the Data field decoder can have high error rates.

After looking at the performance of the Sync field, we look at the performance of the decoding of the overall PPDU. Here we look at the packet error rate (PER) which is the probability that the PPDU was not received correctly. This can be due to a number of factors: the Sync was not detected, the wrong Sync field was classified incorrectly, or the data bits in the Data field were not decoded correctly. Since the PER depends on SNR we typically plot PER versus SNR. The IEEE standard focuses on the 10% PER since that is the PER at which the sensitivity of the receiver is specified.

All of these simulations are performed for both an AWGN channel and for Channel Model D, as well as being done for both the HDR and the LDR.

5.4 PHY Performance: Simulation Results

Here we provide the simulation results focusing on different performance metrics. We begin looking at the performance of the Sync field detector. We look at Sync field detection rate in both an AWGN channel and Channel Model D. We do that for both the HDR and LDR Sync fields. Next, we move onto the Sync field classification error rate. The Sync field detector is required to classify the Sync field as either the HDR or LDR Sync field, and the effectiveness of that classification is studied in an AWGN channel and Channel Model D. Next, we look at the timing error of the Sync field detector, which is required to determine the end of the Sync field, of effectively the beginning of the Data field. For that analysis, we look at the probability mass function (PMF) of the timing error, at the SNR value where the Sync field detection error rate is 10%. Then we move onto the PER curves for the HDR and LDR in both AWGN and Channel Model D. Finally, we look at the PER curves when multiple transmit antennas are used. We see the effect of spatial diversity at the transmitter on those PER curves.

The Task Group spent considerable time evaluating the PHY layer performance, considering different MC-OOK waveform design, for different transmit diversity designs, all under different channel conditions [3–21].

5.4.1 Sync Field Detection Rate

Before going over the PER results for the full PPDU let us look at the Sync detector performance. We will begin with the Sync field detection error rate, which is the probability that the Sync field was not detected.

Figure 5.3 shows the Sync field detection error rate for an AWGN channel, for both the HDR and the LDR.

It is common to focus on the 10% error rate. The SNR at the 10% error rate is around −7.5 dB for the HDR and around −9.5 dB for the LDR. As we will see later, these SNR values are lower than that for the overall PER, since if the Sync field is not detected, clearly the Data field cannot be decoded. This is part of the PPDU design philosophy, that the detection of the Sync field should be more robust than the decoding of the Data field, so that the performance is limited by the Data field performance.

Figure 5.4 shows the Sync field detection error rate for Channel Model D, for both the HDR and the LDR. Here we see that the slope of the curve is much more gradual than that for the AWGN channel, which is due to the effects of the multipath channel. Just like for AWGN channel, the LDR operates at a lower SNR than the HDR. The SNR at the 10% error rate for is around −1.5 dB

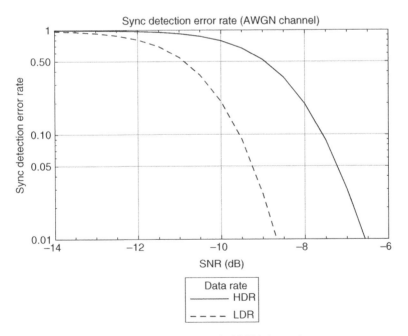

Figure 5.3 Sync field detection error rate in AWGN channel.

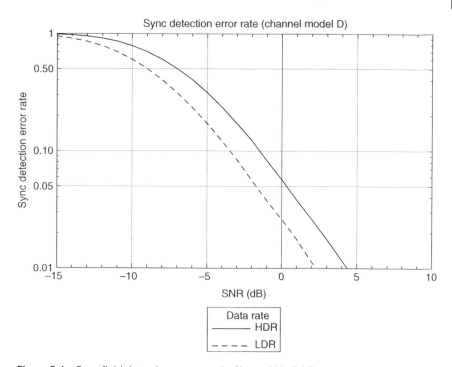

Figure 5.4   Sync field detection error rate in Channel Model D.

for the HDR and around −3.5 dB for the LDR. These SNR values are higher than for the AWGN channel, due to the effects of the multipath channel. Since multipath channels are typical for a wireless local area network, which often operates indoors, there is a strong focus in the IEEE on multipath channels, like Channel Model D.

5.4.2   Sync Field Classification Error Rate

As mentioned earlier, it is possible to detect the wrong Sync field, in other words if the HDR Sync field is transmitted, then it is possible to detect an LDR Sync field. The HDR and LDR Sync fields were designed to be significantly different, so the probability of this happening is low. Also, for all these simulation results, they depend on the details of the Sync detector. Figure 5.5 shows the Sync field classification error rate for AWGN channel. Figure 5.6 shows the Sync field classification error rate for Channel Model D. We see that it is at or below 1%, and at higher SNR levels at which the receiver operates it drops much lower. So, we see that overall, the effect of this classification error rate is very low.

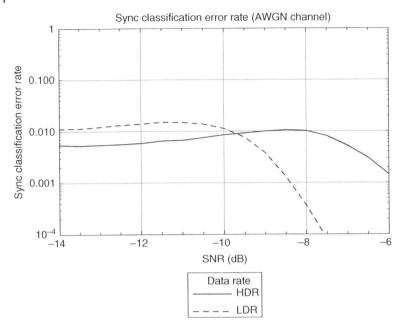

Figure 5.5 Sync field classification error rate in AWGN channel.

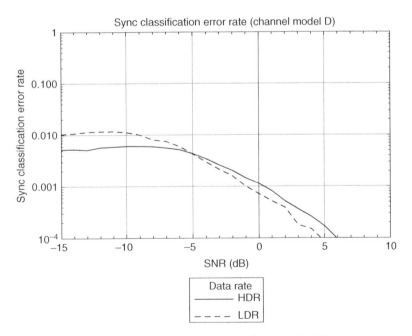

Figure 5.6 Sync field classification error rate in Channel Model D.

5.4.3 Sync Field Timing Error

For the same simulation done for Sync field detection we record the Sync detection timing error for the cases where the Sync field was detected. Clearly, when it was not detected, the timing error does not make sense. Here we need to specify what we mean by the timing error. What we will show is the timing error defined as the difference in the timing under the indicated SNR versus the timing at very high SNR. Since the receiver is sampling at 4 MHz, there will always be some timing error due to the 250 ns sampling resolution. But this 250 ns is small compared to the smallest MC-OOK symbol that we have, which is 2000 ns (2 μs) for the HDR Data field. And for the case of the LDR Data field, the smallest MC-OOK symbol is 4000 ns (4 μs). The timing error can be either positive, negative or zero. Since we are quantizing this to a temporal resolution of 250 ns, the timing error statistics can be shown as a PMF, since the timing error is a discrete random variable. The timing error statistics depend on the SNR, so to plot the PMF of the timing error, we must select an SNR at which to run this simulation. A good choice for that SNR is the SNR at which the missed detection probability is around 10%. That allows us to see if the timing error is a major factor in the performance of the Data field decoding, or if the timing error is not a large effect. Figure 5.7 shows the PMF for the HDR Sync Detector, in an AWGN Channel. We see that in the majority of the cases the timing error is zero, and when it is not zero it is

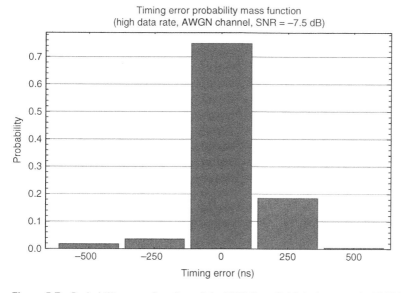

Figure 5.7 Probability mass function of the HDR Sync field timing error in AWGN at SNR = −7.5 dB.

almost always either −250 or +250 ns. So this indicates that the timing error is not a major factor in the performance of the receiver.

Figure 5.7 shows the PMF for the timing error for the HDR Sync field detector in an AWGN channel at an SNR = −7.5 dB, which is where the Sync field detection error rate is around 10%. We see that over 70% of the time, the timing error is zero (at this timing quantization of 250 ns). Then the is around a 20% chance of a timing error of ±250 ns. Finally, there is a low probability of having at timing error of ±500 ns or more.

Figure 5.8 shows the PMF of the timing error for the HDR Sync field in Channel Model D. Here, it is shown at an SNR value of −2 dB, which is approximately the SNR value for a Sync field detection error rate of 10%. As we see the probability of having a timing error of more than ±250 ns is quite low also in this case.

Figure 5.9 shows the PMD for the LDR Sync field timing error in AWGN, at SNR = −9.5 dB

Figure 5.10 shows the PMD for the LDR Sync field timing error in Model D, at SNR = −4 dB. Just like for the HDR the probability of having a timing error of greater than ±250 ns is very low.

Overall, we see that the classification error rate and the timing error are both rather minor issues, so the main issue with the Sync field is detection. If the Sync field is detected then the majority of the time it is classified correctly and has a very small timing error.

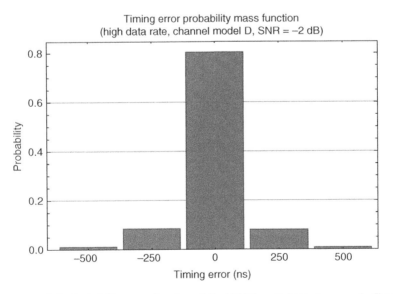

Figure 5.8 Probability mass function of the HDR Sync field timing error in Channel Model D at SNR = −2 dB.

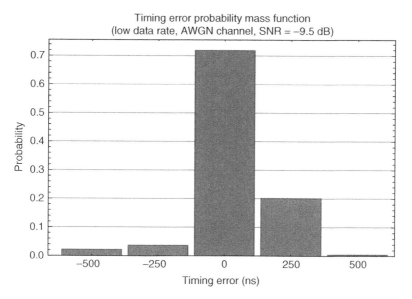

Figure 5.9   Probability mass function of the LDR Sync field timing error in AWGN at SNR = −9.5 dB.

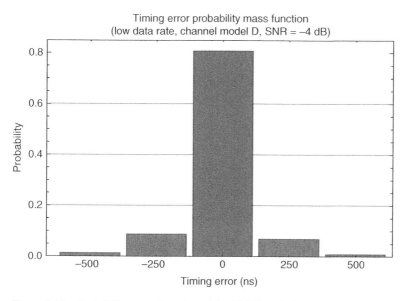

Figure 5.10   Probability mass function of the LDR Sync field timing error in Channel Model D at SNR = −4 dB.

5.4.4 Packet Error Rate

Now we move onto the overall PER performance, for both the HDR and the LDR PPDUs. We expect the PER performance to be somewhat poorer than the Sync detection performance, since if the packet is not detected, it is clearly not decoded.

Figure 5.11 shows the PER, for both the HDR and the LDR, in an AWGN channel. Once again, we see that the LDR operates at a lower SNR than the HDR. The SNR for the HDR at 10% PER is around −4 dB, while the SNR for the LDR at that PER is around −7.5 dB. So, there is approximately 3.5 dB difference between the SNR for the two data rates, in AWGN.

Figure 5.12 shows the PER, of both the HDR and the LDR, in Channel Model D. The SNR for the HDR at 10% PER is around 2 dB, while the SNR for the LDR at that PER is around −1.5 dB. So, there is approximately 3.5 dB difference between the SNR for the two data rates in Channel Model D.

5.4.5 Effects of Transmit Diversity

In all the simulations up until now, we have had only a single transmit antenna. Now we look at the effects of having multiple transmit antennas. It is not

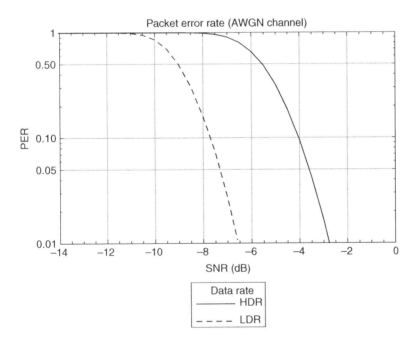

Figure 5.11 HDR and LDR packet error rate in AWGN.

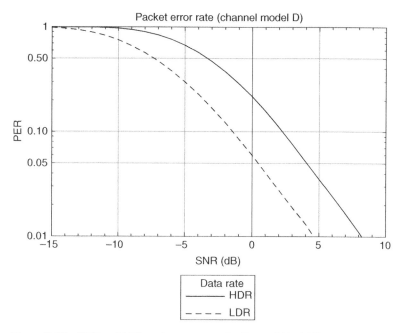

Figure 5.12 HDR and LDR packet error rate in Channel Model D.

uncommon for an IEEE 802.11 AP to have four or more antennas. As described earlier, CSD is applied to both the Sync and Data fields, to minimize the effects of destructive interference at the receiver. The choice of the CSD depends on the number of antennas and the antenna number. Here we use the CSD recommended in Annex AC of the IEEE 802.11ba standard.

Simulations for one, two, four, and eight antennas are run, which provides a good set of simulations, showing the effects of transmit diversity. In all these simulations the total transmit power is kept constant, by scaling the transmit power on each antenna by the inverse of the number of antennas. In practice, if the power level using one antenna is not at the maximum allowed by regulations, then the AP may put out more power when multiple antennas are used. That is because, the power on each antenna may be limited by the choice of the power amplifier, so using more than one antenna may allow for higher overall transmit power. But in our simulations, we scale the transmit power to the same for all antenna configuration, so these simulations do not include any aspects of power gain from multiple antennas

Figure 5.13 shows the HDR PER in AWGN for one, two, four, and eight antennas. We see that in an AWGN channel the PER actually gets a little worse with

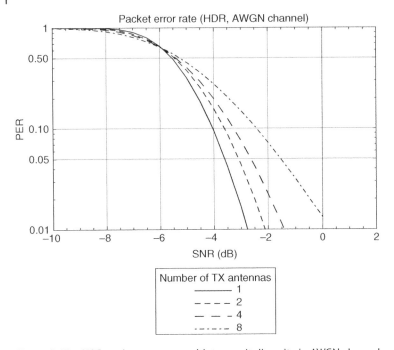

Figure 5.13 HDR packet error rate with transmit diversity in AWGN channel.

additional antennas. This is because there is no diversity between the MISO channels, and there is still some possibility of destructive interference.

Figure 5.14 shows the HDR PER in Channel Model D for one, two, four, and eight antennas. Here we see the benefit of transmit diversity. We see good improvement in performance with multiple transmit antennas due to the diversity of the MISO channels. We see that as we increase from a single antenna to four antennas, we get approximately a 3 dB diversity gain at 10% PER while increasing to eight antennas, we get approximately a 3.5 dB diversity gain.

Figure 5.15 shows the PER for LDR with multiple transmit antennas in an AWGN Channel. Once again, we see for AWGN the PER performance is actually a little worse for multiple TX antennas.

Figure 5.16 shows the PER for LDR with multiple antennas in Channel Model D. Once again we see up to around 3.5 dB gain in PER performance with eight transmit antennas versus a single transmit antenna.

Since real-life channels are typically multipath channel, like Channel Model D, the use of transmit diversity can be very beneficial.

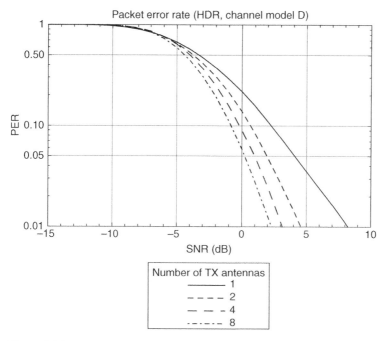

Figure 5.14 HDR packet error rate with transmit diversity in Channel Model D.

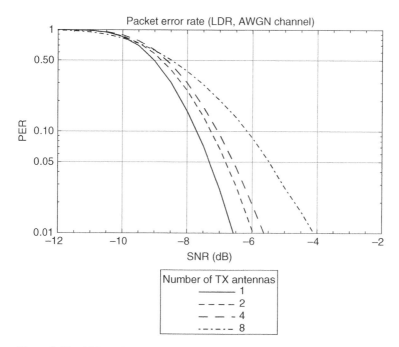

Figure 5.15 LDR packet error rate with transmit diversity in AWGN channel.

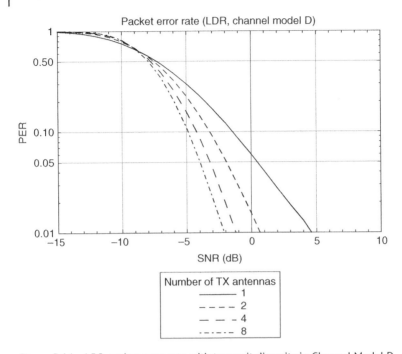

Figure 5.16 LDR packet error rate with transmit diversity in Channel Model D.

5.5 Link Budget Comparison

Recall from Chapter 4 that one of the objectives was for the WUR PHY to have approximately the same range as the MR PHY, which could be the OFDM PHY with a data rate of 6 Mb/s or the even a lower rate non-OFDM 1 Mb/s PHY. Here we will begin by comparing to the 6 Mb/s OFDM PHY and then say a few words about the non-OFDM 1 Mb/s PHY.

There are many factors that affect the link budget for the WUR and the OFDM PHYs. One of the factors includes the transmit power, which in some regulatory domains depends on the transmit bandwidth, which is different for the WUR and the OFDM PHYs. Another factor is the noise figures of the WUR and OFDM receivers. It is expected that in many implementations the noise figure of the WUR receiver is higher than that for the OFDM receiver in order to save power consumption, but that is an implementation choice, so we do not know what the difference is between the noise figure of the two receivers. Here we investigate the required SNR for the WUR, which depends on the channel model and the number of transmit antennas. We assume that the number of transmit antennas for the WUR PHY and the OFDM PHY is the same since the same physical antennas in the AP will be used to transmit the WUR PPDU and the OFDM PPDU. We assume a single receive antenna for this analysis, since it is expected that the STA is an ultra-low-power device. Here we will use the single TX antenna since the required SNR is easily available for the 6 Mb/s OFDM PHY [2]. We focus on Channel Model D, which is an important multipath channel in IEEE 802.11. As you may recall, the SNR for WUR was measured in a 20 MHz channel, just like it is done for the OFDM PHY. This was done to make an easy comparison of the link budge for WUR and OFDM PHYs.

The general formula for the difference in the WUR Path Loss budget and the OFDM Path Loss budget is given as,

$$LB_\Delta = LB_{WUR} - LB_{OFDM}$$

$$LB_\Delta = \left(TX_{WUR} - TX_{OFDM}\right) - \left(SNR_{WUR} - SNR_{OFDM}\right) - \left(NF_{WUR} - NF_{OFDM}\right).$$

Here TX_{WUR} and TX_{OFDM} are the transmit powers for the WUR and the OFDM transmitter respectfully. Recall from Chapter 4 the difference between those transmit power values can vary from zero dB to -7 dB depending on the regulatory domain. Since in some regulatory domains the power spectral density limit can cause the maximum WUR transmit power to be less than the maximum OFDM transmit power.

The SNR values, SNR_{WUR}, and SNR_{OFDM}, are the required SNR values for the WUR and the OFDM PHY to meet the target PER of 10%. This depends on the number of transmit antennas and the channel model.

Finally, NF_{WUR} and NF_{OFDM}, are the noise figures values for the WUR and OFDM PHY, respectively. As mentioned earlier, we expect the noise figure of the WUR receiver to be higher than that for the OFDM receiver.

5.5.1 Comparison to the 6 Mb/s OFDM PHY

Given this framework for the link budget comparison, let us apply it to several cases.

We need to make some implementation assumptions about the noise figure differences between the WUR and the OFDM receivers. We will use the same assumption used by the IEEE engineers, which is that the noise figure for the WUR receiver is 8 dB higher than that for the OFDM receiver. In other words, $NF_{WUR} = NF_{OFDM} + 8$. This higher noise figure in the WUR allows for significant power saving in the WUR receiver.

We will perform this link budget comparison for the case of a single transmit antenna. In that case, we know the required SNR for the 6 Mb/s OFDM PHY in Channel Model D, is approximately 10 dB [2]. This data rate is referred to at the modulation and coding scheme zero (MCS0). From our simulations we see that the required SNR for the WUR receiver in Channel Model D is approximately −1.5 dB.

Let us begin with the case that the transmit power for the WUR and the OFDM PHY is the same, which is reasonable in some regulatory domains. In that case, with the above assumptions, the difference in link budget is, $LB_\Delta = 3.5$ dB. Which means that the WUR link has 3.5 dB more link budget than the OFDM link, and hence should be able to reach as far, and possibly a little farther, than the OFDM PHY.

The next case to consider is that the WUR transmit power is 4 dB lower than the OFDM PHY, which is reasonably in some regulatory domains. In that case the link budget difference is, $LB_\Delta = -0.5$ dB, which means that the link budget for the WUR link and the OFDM link is almost identical.

Finally, we consider the case where the WUR transmit power is 7 dB lower than the OFDM PHY, which is reasonable in some other regulatory domains. In that case the link budget difference is, $LB_\Delta = -3.5$ dB, which means that the link budget for the WUR link has 3.5 dB less link budget than the OFDM link. So, in that case, the WUR link would likely not match the range of the OFDM link. In such a scenario, the implementor may want to lower the WUR receiver noise figure, at the cost of additional power consumption.

5.5.2 Comparison to the 1 Mb/s Non-OFDM PHY

Now let us say a few words about the link budge comparison between the WUR link and the 1 Mb/s PHY link. The 1 Mb/s PHY is not an OFDM PHY, and so we will just make an approximate comparison in the difference in the receiver sensitivity of the 6 Mb/s OFDM PHY and the 1 Mb/s PHY, based entirely on the difference in the data rates. This is reasonable since different implementation is likely to have different receiver sensitivity values for the 6 Mb/s OFDM PHY and the 1 Mb/s PHY. The ratio of the data rates of 6 : 1 corresponds to an 8 dB difference. Hence, we would expect that the 1 Mb/s PHY requires approximately 8 dB lower SNR than is required by the 6 Mb/s OFDM PHY.

This tells us that the WUR link budget is likely to be approximately between 4.5 and 11.5 dB less than the link budget for the 1 Mb/s PHY. This significant difference in link budget indicates that the WUR link is likely to have significantly shorter range than the 1 Mb/s PHY, depending on implementation. One way to reduce the link budget difference is to lower the noise figure of the WUR receiver, at the cost of increased power consumption. This is an implementation decision that depends on the usage model and the regulatory domain in which the WUR is to be deployed.

5.6 Conclusions

In this chapter we have focused on the performance of the WUR system, based on a generic non-coherent receiver. We studied the various issues with Sync field detection including detection error, misclassification error between HDR and LDR Sync fields, and finally, the timing error. We noted that misclassification and timing errors are both small issues compared to Sync detection. We then provided overall PER simulation results for both the HDR and LDR, in both AWGN and Channel Model D. We see that the LDR operates at approximately 3.5 dB lower SNR than does the HDR, in both the HDR and LDR. Finally, we showed the effect of using transmit diversity at the AP. We see that there is some loss in performance AWGN, but in Channel Model D, we get around 3–3.5 dB diversity gain as we increase the number of antennas from one antenna to four or eight antennas. All of these results are for a generic receiver used in the simulations. The results in an actual implementation may differ from these results. One such implementation can be found in [22].

References

1 Azizi, S., Shellhammer, S., and Wilhelmsson, L. (2017). IEEE 802.11 TGba Simulation Scenarios and Evaluation Methodology Document. *IEEE 802.11-17/188r10*.

2 Perahia, E. and Stacey, R. (2013). *Next Generation Wireless LANs: 802.11n and 802.11ac*, 2e. Cambridge University Press.

3 Sundman, D., Wilhelmsson, L., and Lopez, M. (2018). OOK Symbol Design. *IEEE 802.11-18/143r2*.

4 Sundman, D., Wilhelmsson, L., and Lopez, M. (2018). Omni-Directional Multiple Antenna Transmission for WUS. *IEEE 802.11-18/144r2*.

5 Shellhammer, S. and Tian, B. (2018). Simulation on the Effect of OFDM Symbol Design. *IEEE 802.11-18/418r0*.

6 Park, E., Lim, D., and Chun, J. (2018). OOK Waveform Generation Follow-Up. *IEEE 802.11-18/421r1*.

7 Jia, J.J., Gan, M., and Lin, W. (2018). WUR Preamble Sequence Design and Performance Evaluation. *IEEE 802.11-18/435r3*.

8 Sahin, A., Yang, R., and Wang, X. (2018). On OOK Waveform Specification. *IEEE 802.11-18/460r1*.

9 Lopez, M., Sundman, D., and Wilhelmsson, L. (2018). MC-OOK Symbol Design. *IEEE 802.11-18/479r2*.

10 Kristem, V., Azizi, S., and Kenney, T. (2018). 2 us OOK Waveform Generation. *IEEE 802.11-18/492r2*.

11 Kristem, V., Azizi, S., and Kenney, T. (2018). WUR Performance Study with Multiple TX Antennas. *IEEE 802.11-18/493r0*.

12 Lim, D. et al. (2018). Evaluation of WUR Sync Sequence. *IEEE 802.11-18/504r1*.

13 Shellhammer, S. and Yang, R. (2018). MC-OOK 'On' Symbol Generation. *IEEE 802.11-18/584r4*.

14 Kristem, V., Azizi, S., and Kenney, T. (2018). Updated Results on WUR Performance with Multiple TX Antennas. *IEEE 802.11-18/772r0*.

15 Shellhammer, S. and Tian, B. (2018). Multiantenna TX Diversity. *IEEE 802.11-18/773r0*.

16 Lopez, M., Sundman, D., and Wilhelmsson, L. (2018). Omni-Directional Multi-Antenna TX through OFDM Symbol Diversity. *IEEE 802.11-18/881r2*.

17 Kristem, V., Azizi, S., and Kenney, T. (2018). Recommendations on OOK Waveform and CSD Setting. *IEEE 802.11-18/1164r2*.

18 Shellhammer, S. and Tian, B. (2018). Simulations with Recommended Symbols and CSD. *IEEE 802.11-18/1198r0*.

19 Shellhammer, S. and Tian, B. (2018). Comparison of 2 μs MC-OOK Symbols. *IEEE 802.11-18/1199r0*.

20 Shellhammer, S. and Tian, B. (2018). CSD Simulations. *IEEE 802.11-18/1556r0*.

21 Kristem, V., Azizi, S., and Kenney, T. (2018). CSD Recommendations for Example Sequences. *IEEE 802.11-18/1562r0*.

22 Liu, R., Asma Beevi, K.T., and Dorrance, R. (2020). An 802.11ba-based wake-up radio receiver with Wi-Fi transceiver integration. *IEEE Journal of Solid-State Circuits* 55 (5): 1151–1164.

6

Wake-up Radio Medium Access Control

6.1 Introduction

This chapter describes the key functionalities of the medium access control (MAC) layer of IEEE 802.11ba, with a focus on how wake-up radio (WUR) is integrated into and improves network discovery, connectivity, synchronization, and power management (PM). In addition, the chapter describes how frequency division multiple access (FDMA) and protected WUR frames are utilized to improve efficiency and security of transmissions over the WUR. If a station (STA) is equipped with a main radio (MR) and a WUR, and is capable of receiving WUR PPDUs via the WUR radio, and supports the WUR operation, we refer to the STA as a WUR STA in the rest of this book. Similarly, if an access point (AP) is equipped with a MR and a WUR, is capable of transmitting WUR PPDUs via the WUR radio, and supports the WUR operation, we refer to the AP as a WUR AP.

6.2 Network Discovery

6.2.1 General

During network discovery a STA scans the available channels in the frequency bands it supports to discover a suitable AP to which the STA may associate with or eventually roam to, at some point in the future. Traditionally, the STA chooses between two scanning methods:

- **Passive scanning:** The MR of the STA listens on each channel for Beacon frames that are sent periodically by APs
- **Active scanning:** The MR of the STA transmits Probe Request frames and then listens for Probe Response frames sent by APs that received the Probe Request frames

IEEE 802.11ba: Ultra-Low Power Wake-up Radio Standard, First Edition.
Steve Shellhammer, Alfred Asterjadhi, and Yanjun Sun.
© 2023 The Institute of Electrical and Electronics Engineers, Inc.
Published 2023 by John Wiley & Sons, Inc.

Passive scanning generally takes more time since the MR STA must listen and wait for Beacon frames, where the interval between these Beacon frames, i.e. the beacon interval, is typically in the order of a hundred milliseconds. In addition, the STA may miss Beacon frames sent by an AP if it does not wait long enough on a specific channel.

Active scanning instead takes less time since the MR STA actively probes for suitable APs, which in turn send their Probe Response frames relatively quickly. Additionally, the likelihood of discovering an AP with active scanning is higher since APs send their Probe Response frames on demand, shortly after receiving the Probe Request frames. On the other hand, to use active scanning the STA needs to be aware of regulatory restrictions that apply to the channels being probed, and eventually account for the risk of causing probe storms (i.e. the solicitation of a large number of Probe Response frames from APs in range), which impacts the network capacity.

Another factor that has a direct impact on the scan time is the number of channels in the supported frequency bands that need to be scanned by the STA until a suitable AP is discovered since the larger the number of channels to be scanned the longer the time it takes to complete the scanning process.

These increased scan times translate to increased power consumption, especially when the STA uses scanning frequently to perform roaming (e.g. in mobility scenarios and/or dynamically changing wireless channels conditions), and oftentimes this leads to a disruption of wireless services (e.g. STA needs to interrupt data communications with the current AP to scan other channels in search of another AP with which it has a better wireless connection).

Hence, using the WUR of the WUR STA to offload certain scanning functionalities leads to reduced power consumption (since the MR may be turned off while scanning via the WUR), and increased stability of the current wireless services (since the MR may continue the data communications with the current AP while the WUR of the WUR STA scans for other WUR APs in other channels). If the STA obtains preliminary information of the suitable APs that may be operating in the area via the WUR instead of the MR, then it helps avoiding probe storms when a large number of STAs try to send Probe Request frames together on the MR (e.g. a large number of Probe Request frames from the phones of passengers upon a train's arrival to train station). However, the WUR of the WUR STA may only passively scan the channels since the WUR of the WUR STA is only capable of receiving WUR frames, and hence the WUR STA still needs to use the MR to complete the discovery process as described below.

6.2.2 WUR Discovery

A WUR AP that has enabled the WUR discovery service is required to periodically transmit WUR Discovery frames in a selected WUR discovery channel. The WUR discovery channel is recommended to be selected from Channel 1 in the 2.4 GHz frequency band and Channels 40, 44, 149, and 153 in the 5 GHz frequency band. This recommendation encourages the adoption of a common set of WUR discovery channels in most regulatory domains, which helps in reducing scanning latency. This is because, in most cases, the WUR scanning STA would not need to scan all available channels in that frequency band but rather only the channels within the set of recommended WUR discovery channels.

The time interval between the target transmit times of two consecutive WUR Discovery frames is defined as the WUR discovery period. WUR Discovery frames transmitted every WUR discovery period include information that can be used by a scanning WUR STA to determine the basic service set identifier (BSSID) and/or the service set identifier (SSID) being advertised by the transmitting WUR AP, in addition to the operating class and operating channel that is used by the advertised network. Because a WUR PHY protocol data unit (PPDU) has limited length as discussed in Chapter 4, instead of transmitting the complete 48-bit BSSID and the full SSID that could be even longer, the WUR Discovery frames carry a compressed BSSID and a compressed SSID, which provide sufficient information for a WUR STA to determine whether the WUR Discovery frames are associated with a WUR AP the WUR STA is looking for. As detailed in Section 7.4.3, a WUR Discovery frame carries the 12 least significant bits (LSBs) and the 12 most significant bits (MSBs) of the BSSID as the compressed BSSID. The compressed SSID is set to the 16 LSBs of the short SSID, which in turn is a 32-bit value calculated over the full SSID using a hash function defined in the IEEE 802.11 standard.

In general, WUR Discovery frames are expected to be sent using low data rate (LDR) rather than high data rate (HDR) PPDUs. The use of LDR provides longer discovery ranges and increased likelihood of reception of WUR Discovery frames by WUR STAs (for example, not all WUR STAs support the reception of WUR PPDUs with HDR while they are required to support the reception of WUR PPDUs with LDR); however, it also leads to longer PPDU transmit times compared to HDR.

WUR Discovery frames are 10 bytes long, which translates to a WUR PPDU transmission time of 1.436 ms when LDR is used and 0.412 ms when HDR is used. These long PPDU transmit times may require larger values for the WUR discovery period to reduce the impact of the WUR discovery service on the channel utilization, especially when the channel is being utilized by other networks and/or other services from the same network (be it for data transfer using the MR, and/or for other WUR services). For example, if one AP is sending WUR Discovery frames in a channel every 100 ms, then the channel utilization is ~1.4%; however, if 50 APs are sending WUR Discovery frames every 100 ms, then the channel utilization would increase to ~70%. As a solution, an increase of the WUR discovery periods by all these APs, for example from 100 ms to 1 second, would reduce the channel utilization to ~7%. In practice the selection of the WUR discovery period will be a trade-off between channel utilization impact and discovery goals since increasing the WUR discovery period leads to increased scanning latencies for WUR STAs.

The WUR AP may additionally use the MR to advertise that the WUR AP has enabled the WUR discovery service by including a WUR Discovery element in transmitted Beacon frames, and in Probe Response frames that are sent in response to Probe Request frames received from WUR STAs. The included WUR Discovery element provides the WUR discovery operating class and WUR discovery channel, and eventually other discovery-related parameters that are useful for a scanning WUR STA, such as the short SSID, the BSSID, and the WUR discovery period.

The WUR AP may advertise in a similar way that other neighboring WUR APs have enabled their WUR discovery services by including WUR discovery information of these neighboring WUR APs in transmitted WUR Discovery elements. The advertising WUR AP may obtain the WUR discovery information of neighboring WUR APs from WUR Discovery elements that were present in previously received Beacon frames, and Probe Response frames sent by neighboring WUR APs, which in turn may contain WUR discovery information of their own neighboring WUR APs and so on. Hence, a received WUR Discovery element ends up being informative for a WUR scanning STA to eventually narrow down its search space, for example limiting scanning only on the list of WUR discovery channels included in the WUR Discovery element and/or adjusting the scan times to account for the included WUR discovery periods. However, this information is not sufficient to univocally discover a particular WUR AP, be it the transmitting AP or a neighboring AP. This is because the STA may be in the reception range of the MR of the advertising AP but might not be in the reception range of the WUR of the advertising AP (e.g. different transmit powers, path loss, etc.) or of the neighboring APs (likely in a very different location).

A WUR STA that has enabled WUR discovery service performs scanning over a list of WUR discovery channels. The list of WUR discovery channels may be initially limited to the recommended WUR discovery channels that are located in the band(s) supported by the WUR STA (e.g. if the WUR STA only supports operating in the 2.4 GHz band, then WUR scanning is performed only in Channel 1), eventually updated to include any WUR discovery channels that are of interest for the WUR STA and that were obtained from recently received WUR Discovery elements or from past WUR discovery phases. A WUR discovery channel may be considered of interest for a WUR STA if that channel is/was being used to advertise the presence of a network that the WUR STA intends to associate with or intends to discover (e.g. for roaming purposes).

The WUR STA is required to scan a given WUR discovery channel for a minimum amount of time before moving to the next WUR discovery channel in the WUR discovery channel list. Ideally, the minimum channel scan time should be at least equal to the WUR discovery period that is being used by an advertising WUR AP, while in practice it is a parameter that is selected by the WUR STA. If the scanning WUR STA receives a WUR Discovery frame that advertises a network of interest then the WUR STA may switch to the operating channel that is

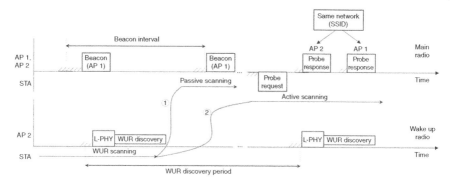

Figure 6.1 Transition from WUR scanning to MR scanning.

being advertised by the WUR Discovery frame to continue the discovery via the MR, or to initiate the association procedure via the MR. Generally, the WUR STA will initially perform active scanning via the MR to ensure that the WUR STA is not only within reception range of the WUR but also in the reception range and the transmission range of the MR of a WUR AP that belongs to that same network and then initiate the association procedure. Figure 6.1 shows examples of a WUR STA that transitions, after receiving a WUR Discovery frame from WUR AP 2, from WUR scanning to passive scanning on the MR in the first example, and that transitions from WUR scanning to active scanning on the MR in the second example. In both examples, the WUR STA turns on the MR and switches to the operating channel that is indicated in the received WUR Discovery frame.

In the first example, the WUR STA's MR initiates passive scanning in that channel. During passive scanning the WUR STA's MR receives a Beacon frame from AP 1, which has a BSSID that is not the same as the BSSID indicated by the BSSID of the WUR Discovery frame, although AP 1 belongs to the same network (i.e. the short SSID derived from the SSID in the Beacon matches the short SSID of the WUR Discovery frame). Additionally, AP 1 does not support the functionalities that are expected by the STA (for example, WUR support). Hence, the STA needs to either continue passive scanning (not shown in the figure) or enable active scanning to discover AP 2.

In the second example, the WUR STA's MR initiates active scanning by transmitting a broadcast Probe Request frame in that channel and subsequently waits to receive Probe Response frames that are expected to be generated by APs that have received the Probe Request frame (AP 1 and AP 2 in our example). Each Probe Response frame includes the BSSID of the transmitting AP, the SSID, and a list of all the functionalities that are supported by each AP, with WUR AP 2 indicating WUR support, which is of interest to the WUR scanning STA. After receiving the Probe Response sent by WUR AP 2, the WUR STA may immediately initiate the association process with AP 2.

6.3   Connectivity and Synchronization

6.3.1   General

A WUR AP sends WUR Beacon frames periodically to help a WUR STA to detect if the WUR STA is in range of the WUR AP, to maintain time synchronization.

6.3.2   WUR Beacon Frame Generation

A WUR AP that has enabled the WUR service is required to periodically transmit WUR Beacon frames in a specific channel, which is defined as the WUR primary channel. The WUR beacon period is the time interval between two consecutive target WUR beacon transmit times (TWBTTs).

The WUR primary channel is recommended to be located in the same frequency band as the primary channel of the MR, noting that the WUR primary channel cannot be located in a channel that is primarily located for radiolocation radar systems and the AP has in-service monitoring requirements for 50–100 µs radar pulses.

The WUR beacon period is implementation specific and recommended to be chosen based on incurring overhead, timing correction requirements, and other considerations (e.g. a WUR AP may choose a WUR beacon period that would satisfy the requirements of a WUR STA that requests the WUR AP to generate keepalive WUR frames with a certain periodicity (see Section 7.2.3.3)).

The WUR AP provides all the parameters that control the schedule of WUR beacons (i.e. where and when these frames are generated) in the WUR Operation element, which is included in specific Management frames (see Section 7.3) transmitted by the WUR AP via the MR. Figure 6.2 illustrates that the parameters are announced in Beacon frames sent by the WUR AP via the MR. These parameters, namely the WUR primary channel, the WUR beacon period, and the offset of the first TWBTT, are obtained from the WUR Operating Class and WUR Channel, WUR Beacon Period, and Offset Of TWBTT subfields, respectively of the WUR Operation element (see Section 7.2.3.2).

The WUR AP schedules for transmitting WUR Beacon frames via the WUR at each TWBTT, as illustrated in Figure 6.2 using the parameters announced in the WUR Operation element. The WUR AP ensures that the WUR Beacon frames are transmitted in a WUR PPDU with a rate (LDR or HDR) that is supported by all associated WUR STAs in WUR mode. To help a WUR STA synchronize with the WUR AP, the WUR Beacon frames contain the partial timing synchronization function (TSF) in the type-dependent control field of the WUR Beacon frame. The partial TSF is chosen instead of the complete TSF, which is 8 octets, to provide enough information for time synchronization while minimizing the length of the WUR Beacon frame. The partial TSF is equal to the value of the TSF timer [5 : 16] at the time that the first OOK symbol containing the first bit of the TD control

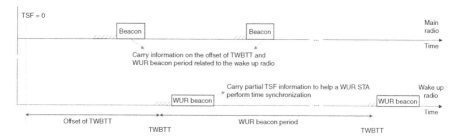

Figure 6.2 Maintaining WUR connectivity and synchronization with WUR Beacon frames.

field is transmitted plus any delays through local MAC/PHY interfaces, etc. The STA on the other hand can obtain the full TSF timer from Beacon frames, and Probe Responses that the AP sends via the MR.

There are also some additional rules the AP needs to follow when transmitting the WUR Beacon frames.

One rule is for a WUR AP that supports the multiple BSSID capability defined in IEEE 802.11ax, with which the AP can indicate multiple 802.11 networks on the same channel in a single Beacon or Probe Response frame using the Multiple BSSID element. Each of these networks is referred to as a BSS in IEEE 802.11 and has a unique 48-bit BSSID and a corresponding SSID. For example, the AP may indicate one BSS which SSID is set to "home" and another BSS which SSID is set to "guest" on the same channel and advertising them using a single Beacon frame. The BSSID of the BSS that is used to generate the Beacon frames that contain the indicates the Multiple BSSID element is called the transmitted BSSID (e.g. the BSSID of the BSS named "home"), and the BSSID(s) indicated by the Multiple BSSID element are known as the non-transmitted BSSIDs (e.g. the BSSID of the BSS named "guest"). In this setting, we refer to that a STA associated with the AP as the STA associated with any of the transmitted BSSID or non-transmitted BSSIDs. IEEE 802.11ba has a rule that such an AP ensures that WUR Beacon frames are only transmitted by the transmitted BSSID.

In general a WUR Beacon frame needs to compete for the medium along with other pending frames that are scheduled for transmission via the MR at a given TWBTT. In this case, the transmission of the WUR Beacon frame has lower priority than scheduled Beacon frames but higher priority than any other pending frames that are scheduled for transmission via the MR.

There are cases when the WUR Beacon frames from the WUR AP are not needed. For example, there might be no WUR STA associated with the AP (or any of the WUR APs in the multiple BSSID set with the AP). In addition, the AP may decide not to provide WUR PM service to any associated WUR STA. Lastly, all WUR STAs associated with the WUR AP may be in an Active mode. In order to improve airtime efficiency in these cases, the WUR AP is allowed to not transmit WUR Beacon frames.

6.3.3   WUR Beacon Frame Processing

A WUR STA adopts the WUR beacon period when joining the BSS and expects to receive WUR Beacon frames from the WUR AP every WUR beacon period. The STA obtains the parameters that are being used by the WUR AP for scheduling WUR Beacon frames (i.e. WUR primary channel, WUR beacon period, and offset of the first TWBTT) from the WUR Operation element included in the most recent Management frame transmitted by the AP via the MR, as illustrated in Figure 6.2.

The WUR STA uses these parameters to determine the actual wake-up pattern that the STA will follow to receive WUR Beacon frames when operating in WUR mode. The actual wake-up pattern depends on several factors, such as power consumption budget (the more frequent the wake-up pattern the higher the power consumption), TSF clock accuracy (WUR STAs are required to have a TSF accuracy of ±100 ppm which means that the clock drift must be less than 100 μs every second), and connectivity resilience. If the STA needs more frequent WUR Beacon frames, then the STA may request the AP to generate keep-alive frames (see Section 7.2.3).

The WUR STA then ensures to be awake at TWBTTs and waits to receive WUR Beacon frames sent by the WUR AP. During this time, the STA discards any other WUR frames that are not WUR Beacon frames.

The successful reception of a WUR Beacon frame with a matching transmitter ID indicates to the WUR STA that it is currently within range of the WUR AP. If the WUR STA does not receive a WUR Beacon frame with a matching transmitter ID for a certain period of time, then the WUR STA should either start WUR scanning or wake its MR and probe the AP to check if it is still within range of the AP via the MR. If the WUR STA is not able to find the WUR AP either via WUR scanning or via the MR then the WUR STA assumes the connection has been lost. If the WUR STA chooses to not receive any WUR Beacon frames, then it might not be able to determine that it is out of range in a timely manner.

A received WUR Beacon frame additionally helps the WUR STA to update its TSF timer without the need of waking the MR. This helps to save power at the MR. The TSF timer of the STA is updated using the partial TSF timestamp that is included in the received WUR Beacon frame after being adjusted by the internal delays incurred by the WUR STA with the following steps:

- Create a temporary timestamp by concatenating the received partial TSF timestamp with 5 bits containing an implementation-specific value that represents the assumed value of bit positions 0–4 of the temporary timestamp.
- Add an amount equal to the receiving WUR non-AP STA's delay through its local PHY components plus the time since the first bit of the Partial TSF field was received at the MAC/PHY interface to the temporary timestamp.
- The adjusted value of the received partial TSF timestamp is set as the value of bit positions 5–16 of the temporary timestamp.

The WUR STA obtains an adjusted value after the procedure above, denoted as adjusted time (AT) here.

AT may not be an accurate estimate of the TSF at the AP, if the mutual clock drift between the AP and the STA is too large. For example, the STA may fail to decode some WUR Beacon frames due to interference, resulting in a mutual drift that is beyond the range that the partial TSF in the WUR Beacon frame can cover. To account for such a rollover, the STA performs the following as mitigation. If the MSB of AT is not equal to B16 of the local TSF timer (denoted as LT), then B17–B63 of LT need to be adjusted as follows:

- The value shall be increased by one unit (modulo 2^{47}) if $LT[5:16] > AT$ and $LT[5:16] > ((AT + 2^{11}) \ (\text{modulo } 2^{12}))$.
- The value shall be decreased by one unit (modulo 2^{47}) if $LT[5:16] < AT$ and $LT[5:16] < ((AT - 2^{11}) \ (\text{modulo } 2^{12}))$.

where $LT[5:16]$ denotes the value of B5–B16 of the local TSF timer.

Lastly, the B5–B16 of the WUR STA's local TSF timer is set to the adjusted value of the received partial TSF timestamp.

6.4 Power Management

6.4.1 General

IEEE 802.11ba has defined new frames and procedures for both the MR and the WUR to achieve both low latency and lower power consumption in an IEEE 802.11 network.

6.4.1.1 MR Power Management

As briefly described in Section 2.3.7, the MR of a STA can be in one of the following two power states:

- Awake: the MR of the STA is fully powered.
- Doze: The MR of the STA is not able to transmit or receive and consumes very low-power.

The STA relies on the MR PM to determine how to transition between these two power states. The STA can be in one of the two PM modes:

- Active Mode (AM): The STA remains in the awake state and can receive and transmit frames at any time.
- Power Save Mode (PSM): The STA enters the awake state to receive and transmit frames and remains in the doze state otherwise.

The STA is required to inform the AP that it is changing the PM mode by completing a successful frame exchange with the AP, where the frame exchange consists of a frame sent by the STA and an acknowledgment frame sent by the AP. The STA indicates the PM mode to which it is transitioning in the Power Management subfield of the Frame Control field of the frame that the STA sends to the AP. The STA is expected to transition to the newly indicated PM mode immediately after the reception of the acknowledgment that is sent by the AP. This protocol ensures that at all times the AP knows the PM mode of any associated STA. This is because the AP is required to buffer frames that are pending for the STA that is in a PS mode until the STA has transitioned to the awake state.

IEEE 802.11 has defined multiple ways for a legacy AP to notify the STAs of buffered unicast and multicast/broadcast frames in Beacon frames. The presence of Buffered unicast frame(s) for a STA is indicated by the TIM element in every Beacon frame as described in Section 2.3.7.

On the other hand, the presence of buffered multicast/broadcast frames is only indicated in a Delivery TIM (DTIM) element of Beacon frames. The DTIM element is a variation of the TIM element and is included in every Beacon frame or in a subset of Beacon frames, depending on the settings of the DTIM Period field of the TIM element. For example, a DTIM period field of 1 indicates that the TIM element in every Beacon frame also serves as a DTIM element; a DTIM period field of 2 indicates that the TIM element in every *other* Beacon frame serves as a DTIM element. Essentially, the DTIM period field determines how often the DTIM element is included in Beacon frames. If a DTIM element is included in a Beacon frame and there is any buffered multicast/broadcast frame, the Traffic Indicator subfield in the Bitmap Control field of the DTIM element is set to 1. To prioritize the delivery of these buffered multicast/broadcast frames, after the Beacon frame containing this DTIM element, the AP transmits them before transmitting any unicast frames.

Besides the presence of buffered frames at the AP, a STA also needs to learn of any critical update on the operation parameters of the BSS. For example, the AP may change the operating channel width or move to a different operating channel due to environment changes. In such case, the STA needs to learn of the change within a short time so as to avoid disrupting the connection to the AP. In legacy 802.11, the MR of the STA has to transition to the Awake state periodically to decode Beacon frames or other management frames (e.g. TIM frames) to learn the change, which is a costly operation in terms of energy consumption.

6.4.1.2 WUR Power Management

The WUR PM defined in 802.11ba helps a WUR STA learn the presence of buffered frames or the critical update using the WUR instead of relying on waking up the MR periodically. As the WUR has much lower power consumption than the

MR, the WUR PM not only saves more power but also reduces latency for the STA to retrieve the frames or update.

The PM defined in 802.11ba expands the legacy PM state machine illustrated in Chapter 2 by including the state of the WUR. A WUR AP may provide WUR PM service to its associated WUR STAs. A WUR STA that is using WUR PM service can be either in WUR mode or in WUR mode suspend. Once in the WUR mode, the STA can take advantage of the WUR to achieve lower power consumption and latency. The overall PM state machine with the WUR mode is shown in Figure 6.3.

As shown in Figure 6.3, the state machine for the MR, shown at the top of the figure, remains exactly the same as that in the legacy 802.11 as shown previously in Figure 2.9. This is because 802.11ba is designed to be backward compatible. The 802.11ba expansion, indicated as the WUR power state in the WUR mode, is shown at the bottom of the figure.

In the WUR mode, the WUR STA can be in the WUR awake state or in the WUR doze state. The WUR STA can simply stay in the WUR awake state all the time, which will allow the WUR AP to wake up the WUR STA via the WUR at any time to minimize wake-up latency. To further reduce the average power consumption of the WUR, 802.11ba also allows the WUR STA to transition between the WUR awake state and the WUR doze state based on transition events. These events can be scheduled (see Duty cycle operation in Section 6.4.3) or unsolicited at any time.

While in the WUR awake state, if the STA receives a WUR Wake-up frame from the AP about buffered frames or critical update of the BSS parameter, then the

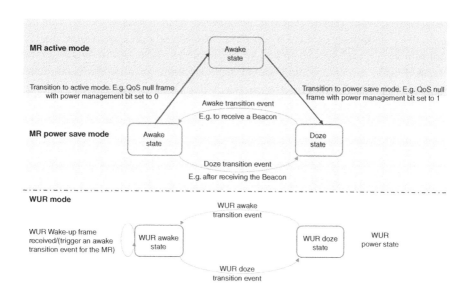

Figure 6.3   IEEE 802.11ba power management.

WUR will trigger an awake transition event for the MR. From this point, the MR of WUR STA follows the PS operation that the WUR STA has negotiated with the WUR AP to get into the Awake state and to retrieve the buffered frame or critical update.

With the awake transition event triggered by the WUR, the MR of a WUR STA does not have to transition to the Awake state periodically as in legacy 802.11 to learn buffered frames or critical update anymore. By eliminating the periodic transitions to the Awake state on the MR, the energy consumption of the MR will be future reduced. In addition, it can reduce the latency by waking up the MR via the WUR. This is because the WUR Wake-up frame may be received by the WUR STA sooner than the next scheduled notification on the MR (e.g. the TIM/DTIM element in the next Beacon frame), if the WUR DC is configured appropriately.

The remaining sections of this chapter will focus on detailed descriptions regarding the available protocols and functionalities that are provided to the WUR STA when in WUR mode.

6.4.2 WUR Modes

A WUR STA that intends to enable the WUR PM service needs to successfully complete WUR mode setup with the associated WUR AP. After enabling the WUR PM service, the STA may dynamically change the WUR status, receive updates from the WUR AP on the operational parameters relevant to the WUR PM service, and finally tear down the WUR PM service when it is not needed anymore. The WUR status can be either *WUR mode* or *WUR mode suspend*. The WUR mode has been discussed above, whereas the WUR mode suspend indicates that the WUR is not in use, although the negotiated WUR parameters between the WUR AP and the WUR non-AP STA are maintained. In this section, we first discuss in detail each of these functionalities, then provide a flow chart that gives the overall picture of these functionalities and WUR state transitions, and finally conclude with a practical example depicting the frame exchanges that are involved in enabling, suspending, updating, and tearing down the WUR PM service.

6.4.2.1 WUR Mode Setup

WUR mode setup is required for enabling the WUR PM service between a WUR STA and a WUR AP and negotiating the corresponding WUR parameters. WUR mode setup can be performed during the (re)association phase by including the WUR Mode element in exchanged (Re)Association Request frames, and (Re)Association Response frames exchanged between the WUR STA and the WUR AP provided that the STA is aware of the WUR Capabilities of the AP and of the parameters that the WUR AP advertises in transmitted WUR Operation elements. For example, the WUR STA can obtain these elements during scanning, as described in Section 6.2. WUR mode setup can also be performed post association by including the WUR Mode element in exchanged WUR Mode Setup frames between the WUR STA and the WUR AP. Performing WUR mode setup during (re)association may reduce over-the-air frame exchanges, since there is no need to exchange additional WUR Mode Setup frames, however it is not the preferred choice from a security perspective. WUR mode setup should rather be performed after association, or more precisely after authentication, because this way the WUR PM service negotiation can be performed with protected WUR mode setup frames, which enable the exchange of WUR parameters in a secure way.

For the remainder of this book, we will refer to frames that contain a WUR Mode element and that are sent by a WUR STA as "WUR Mode Request frames," and to frames that contain a WUR Mode element and that are sent by a WUR AP as "WUR Mode Response frames."

A WUR STA requests the enablement of the WUR PM service by transmitting a WUR Mode Request frame to the AP. The frame is successfully delivered if the WUR STA receives in response from the WUR AP an acknowledgment (Ack) frame. The WUR STA may retransmit the WUR Mode Request frame until it is successfully delivered. The WUR Mode element of the WUR Mode Request frame indicates:

- The WUR status that the STA expects to have after the WUR mode setup is successful. The Action Type field indicates *Enter WUR Mode Request* if the initial WUR status is expected to be WUR mode and indicates *Enter Mode Suspend Request* if the initial WUR status is expected to be WUR mode suspend.
- A list of parameters, including proposed WUR parameters, that the WUR STA expects to use when in WUR mode such as those described in detail in Sections 6.4.4–6.4.6 and 6.5.

The WUR AP responds to a WUR Mode Request frame sent from a WUR STA by transmitting a WUR Mode Response frame to the WUR STA. Similarly, the frame is successfully delivered if the WUR AP receives in response from the WUR STA an Ack frame, and the WUR AP may retransmit the WUR Mode Response frame until it is successfully delivered. The WUR Mode element of the WUR Mode Response frame indicates:

- A reason when the request is denied, regardless of whether the request is accepted or denied. If the request is accepted, then the WUR mode setup is successful.
- The WUR status of the STA if the request is accepted. The Action Type field indicates *Enter WUR Mode Response* if the requested WUR status was WUR mode and indicates *Enter WUR Mode Suspend Response* if the requested WUR status was WUR mode suspend.
- A list of WUR parameters that the STA needs to use when operating in WUR mode. These parameters are described in detail in such as those described in detail in Sections 6.4.4–6.4.6 and 6.5.

Following a successful WUR mode setup, the WUR AP and the WUR STA maintain a WUR status for the WUR STA, which is either WUR mode or WUR mode suspend.

If the WUR STA's WUR status is WUR mode, then the WUR AP and WUR STA maintain an up-to-date list of the negotiated WUR parameters between the two. The WUR AP schedules, during specific WUR DC service periods that are negotiated as described in Section 6.4.3, the transmission of specific WUR frames intended for the WUR STA. Similarly, the WUR STA's WUR needs to be in Awake state, during specific WUR DC periods that are negotiated, for the reception of specific WUR frames intended for the STA and eventually perform certain actions in response to the reception.

If the WUR STA's WUR status is WUR mode suspend, then the WUR AP and WUR STA maintain an up-to-date list of the negotiated parameters between them. However, neither the WUR AP nor the WUR STA follows the negotiated schedules and protocols described in the abovementioned sections for as long as the STA's WUR status remains WUR mode suspend.

6.4.2.2   WUR Mode Update

The WUR mode update enables the WUR AP to update the operational parameters, including WUR parameters, of a WUR STA that has an established WUR PM service with the WUR AP. WUR mode update can be performed after association and only after a successful WUR mode setup between the WUR AP and the WUR STA.

A WUR AP updates the WUR parameters by successfully transmitting a WUR Mode Response frame to the WUR STA. The WUR Mode Response frame contains:

- A WUR Mode element that contains any updates to WUR parameters that are determined by the AP and that are specific for this particular STA. The Action Type field indicates *Enter WUR Mode Response* if the WUR status of the STA is WUR mode and indicates *Enter WUR Mode Suspend Response* if the WUR status of the STA is WUR mode suspend.
- An optionally present WUR Operation element that contains updates, if any, to the WUR parameters that are applicable to all WUR STAs that have established WUR PM service with the AP, which are described in Section 7.2.3.
- An optionally present WUR PN Update element that contains updates, if any, to security-related WUR parameters, which are described in Section 7.2.3.

The WUR STA is required to update the locally stored WUR parameters to the WUR parameters included in the received WUR Mode Response frame. The updated WUR parameters are expected to take effect immediately after the WUR STA responds with an Ack frame to the received WUR Mode Response frame.

6.4.2.3   WUR Mode Suspend and Resume

WUR mode suspend, and resume enables a WUR STA to dynamically transition from WUR mode to WUR mode suspend, and vice versa.

WUR mode suspend, i.e. WUR mode to WUR mode suspend transition, is achieved by successfully transmitting a WUR Mode Request frame to the WUR AP when the WUR STA's WUR status is WUR mode. The Action Type field of the WUR Mode element in the WUR Mode Request frame indicates *Enter WUR Mode Suspend.*

WUR mode resume, i.e. WUR mode suspend to WUR mode transition, is achieved by successfully transmitting a WUR Mode Request frame to the WUR AP when the WUR STA's WUR status is WUR mode suspend. The Action Type field of the WUR Mode element in the WUR Mode Request frame indicates *Enter WUR Mode.*

The WUR AP is required to update the WUR status of the WUR STA to the WUR status indicated in the most recently received WUR Mode Request frame. The update of the WUR status is expected to occur immediately after the WUR AP responds with an Ack frame to the received WUR Mode Request frame.

6.4.2.4   WUR Mode Teardown

WUR mode teardown is required for tearing down an established WUR PM service between a WUR STA and a WUR AP, which helps freeing the resources at the WUR AP and WUR STA that were being used for providing WUR PM service to the WUR STA. WUR mode teardown can only be performed after a successful WUR mode setup between the WUR AP and the WUR STA. The WUR AP or WUR STA may tear down the WUR PM service by successfully transmitting a WUR Mode Teardown frame to the WUR STA or WUR AP, respectively.

A flow chart that summarizes all the different processes needed to transition in, out, and between different WUR modes is shown in Figure 6.4.

An example that shows the frame exchange sequences involved in enabling the WUR PM service of a WUR STA, transitioning from WUR mode to WUR mode suspend, updating WUR parameters, and tearing down the WUR PM service is shown in Figure 6.5.

The WUR STA enters the WUR mode after the successful completion of WUR mode setup and remains in WUR mode for as long as this mode of operation aligns with the patterns of its downlink traffic. At a certain point, the pattern of the downlink traffic changes to a degree that there is no alignment with the STA's WUR mode in which case the WUR STA notifies the AP that the WUR STA has entered WUR mode suspend. The WUR STA's WUR transitions to WUR doze state, while in WUR mode suspend, and all interactions with the WUR AP will occur via the MR from now on. At this point in time, there is an expectation by the WUR STA that operation in WUR mode will resume, and hence the WUR STA

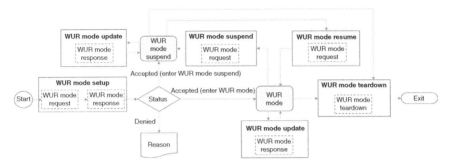

Figure 6.4 Flowchart diagram of WUR modes.

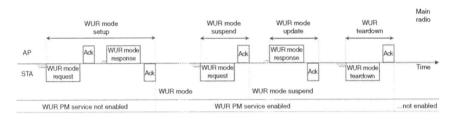

Figure 6.5 Example with different WUR mode functionalities.

suspends rather than tears down the WUR mode. Later on, the WUR AP notifies the WUR STA of an update to the WUR parameters, for example that the WUR channel is now changing from the 2.4 GHz band to the 5 GHz band. The WUR STA then tears down the WUR mode because this particular WUR STA does not support WUR operation in the 5 GHz band.

6.4.3 Duty Cycle Operation

It is not necessary to keep the WUR in awake state at all times, especially when there is sparse traffic, so 802.11ba has also introduced duty-cycled operation to further reduce the power consumption of the WUR.

A WUR STA that keeps the WUR in awake state at all times will continuously consume power, P_{WUR}, for the duration of the time that the STA operates in WUR mode, T_{WUR}, which leads to an energy consumption by the WUR of the WUR STA that is $E_{WUR} = P_{WUR} \times T_{WUR}$. As discussed in Chapter 3, P_{WUR} is significantly lower than the power consumed by the MR of the STA, P_{MR} (i.e. $P_{WUR} << P_{MR}$). Hence, if T_{WUR} is comparable to the duration of time that the WUR STA would need to maintain the MR in the awake state, T_{MR}, (i.e. $T_{WUR} \approx T_{MR}$), then the energy consumption will decrease proportionally with the power consumption (i.e. $E_{WUR} << E_{MR}$).

However, things start to look differently when we consider that the MR has power save protocols, described briefly in the previous sections, which reduce the amount of time the STA needs to keep the MR in awake state, T_{MR}, which reduces the energy consumption of the MR. Hence, keeping the WUR in awake state at all times may lead to higher energy consumption when compared to the energy consumption of an MR that is using power save protocols.

Consider for example a case where the WUR STA has kept the WUR turned on for a total of 10 seconds, while during this same amount of time the MR has been turned on only for 0.1 second because the MR has scheduled periodic wake times (using TWT with 1 ms service periods and intervals of 100 ms). Assuming for example a P_{WUR} of 1 mW and a P_{MR} of 50 mW, the total energy consumption of the WUR will be twice as large as that of the total energy consumption of the MR. Hence, the WUR would not help reducing the overall energy consumption of the STA. For this reason, WUR duty cycle (DC) operation is defined, which adds another degree of freedom for further reducing the energy consumption of the WUR.

The duty cycle, D_C, is defined as the ratio between the time that the WUR of a STA is expected to be in awake state, T_{ON}, and the total WUR operation time T_{TOT}, where T_{TOT} is equal to $T_{ON} + T_{OFF}$, with T_{OFF} being the time that the WUR of the STA is expected to be in doze state. Hence, D_C can be expressed as:

$$D_c = \frac{T_{ON}}{T_{TOT}}$$

WUR DC operation is negotiated between a STA and an AP during WUR mode setup and is governed by the three WUR parameters defined below.

6.4.3.1 WUR Duty Cycle Period

The WUR DC period, T_{DCP}, determines the periodicity with which the WUR STA intends to wake up the WUR. The value of this parameter is selected by the WUR STA because it has a direct impact on the ability of the WUR STA to satisfy certain QoS requirements (such as latency, reliability, etc.) and at the same time on the power consumption of the WUR of the WUR STA. However, the unit for this parameter is provided by the WUR AP in the DC Period Unit field of the WUR Operation element that the WUR AP advertises to the BSS. This way the WUR AP ensures that there is a minimum common denominator for the WUR DC periods that are selected by the associated WUR STAs, which helps in simplifying the scheduling algorithms of the WUR AP (e.g. use a time division multiple access (TDMA) scheduler with discrete start time alignment of the time slots). The WUR DC period is provided by the WUR STA in the WUR DC Period field of the WUR Mode element included in the WUR mode request.

6.4.3.2 WUR Duty Cycle Service Period

The WUR DC service period (SP), T_{DCSP}, determines how long the WUR STA expects to keep the WUR awake for each WUR DC period. This parameter has a similar impact on the ability of the WUR STA to satisfy the above-mentioned QoS requirements and on the power consumption of the WUR of the STA. Hence, similar to the WUR DC period, the value of this parameter is also selected by the WUR STA provided that the WUR DC service period does not exceed the WUR DC period. In addition, along the same similar lines, the WUR AP provides a minimum wake duration in the Minimum Wake Duration field of the WUR Operation element that the WUR AP advertises to the BSS. This ensures that there is a minimum amount of time that all WUR STAs remain awake, which in turn helps in simplifying the WUR frame generation algorithms of the WUR AP (e.g. ensure that the amount of time that the WUR STAs remain awake is greater than the amount of time needed for the WUR AP to contend for the medium and send a WUR frame to the WUR STA (possibly along with other STAs) as described in detail in subsequent sections. The WUR DC service period is provided by the WUR STA in the DC Service Period field of the WUR Mode element that is included in the WUR mode request.

6.4.3.3 WUR Duty Cycle Start Time

The WUR DC start time, T_{DCST}, indicates the start time of the first WUR DC service period (i.e. when the WUR DC operation is expected to start for the WUR STA) and is selected by the WUR AP. This gives the WUR AP the flexibility to align the start times of the WUR DC service periods with the start times of existing WUR DC service periods that the WUR AP has negotiated with other WUR STAs. The start time alignment simplifies the WUR AP's scheduler and is only possible for WUR DC service periods that have a minimum common denominator between the corresponding WUR DC periods.

After successfully completing the WUR mode setup the WUR STA is required to be in WUR awake state for the duration of each negotiated WUR DC service period unless the MR of the WUR STA is either already in the awake state or has been instructed by the WUR AP with Wake-up frames during a particular WUR DC service period to transition the MR to awake state. These exceptions arise from the fact that the AP is not expected to interact with the STA via the WUR while the MR of the STA is in awake state, allowing the STA to further save power by turning off the WUR during WUR DC service periods of no expected WUR activity.

Hence, the WUR STA may decide to be in WUR doze state for the full duration of a WUR DC service period during which the MR of the WUR STA happens to be in awake state, T_{OFF}^{FSP}, for the duration of the remainder of a WUR DC service period after having received the instruction by the WUR AP to transition the MR to awake state $T_{\mathrm{OFF}}^{\mathrm{PSP}}$, and last but not least outside of negotiated WUR DC service periods, $T_{\mathrm{OFF}}^{\mathrm{OSP}}$. This leads to a total duration of time during which the WUR STA can remain in doze state:

$$T_{\mathrm{OFF}} = \sum_{i=1}^{M} T_{\mathrm{OFF},i}^{\mathrm{OSP}} + \sum_{j=1}^{N} T_{\mathrm{OFF},j}^{\mathrm{FSP}} + \sum_{k=1}^{P} T_{\mathrm{OFF},k}^{\mathrm{PSP}}$$

where each time duration $T_{\mathrm{OFF},k}^{\mathrm{PSP}}$, with k in $\{1, \ldots, P\}$, depends on the instant of time that the WUR STA receives the instruction sent by the AP to transition the MR to awake state (see Section 6.4.4).

In Figure 6.6 we provide an example of a WUR STA 1 and a WUR AP setting up WUR DC operation. In this example the WUR STA 1 selects and provides to the WUR AP in the WUR mode request the WUR DC service period, T_{DCSP}, and the WUR DC period, T_{DCP}, subject to the restrictions described above. The WUR AP then responds with a WUR mode response which in addition to accepting the WUR mode also provides the WUR DC start time, T_{DCST}, at which the WUR STA 1 is expected to start its WUR PM service. As it can be seen in the figure the WUR AP selects a T_{DCST} for the WUR STA 1 that aligns the WUR STA's WUR DC schedule with the WUR DC schedules that the WUR AP has previously negotiated with WUR STAs 2, 3, 4. It is worth noting that while the start times of the service periods coincide, not all WUR STAs have the same SP durations nor the same WUR DC periodicities, although the WUR DC periodicities are multiples of a common WUR DC periodicity, which is that of WUR STA 2. After receiving the WUR mode response the WUR STA 1 can

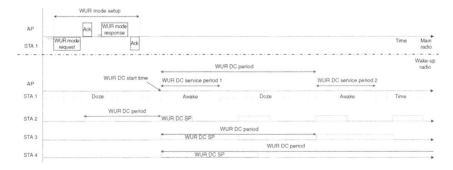

Figure 6.6 WUR duty cycle operation.

determine the start time of any WUR DC service period, n, as $T_{DCST, n} = T_{DCST} + (n-1) \times T_{DCP}$.

In this section, we described the WUR DC operation, which enables the WUR STA to negotiate with the WUR AP periodic wake times and corresponding service periods during which the WUR AP can send frames that wake up the WUR STA. In the following sections, we will focus on describing the different protocols that are available at the WUR AP to cause the WUR STA to wake up the MR.

6.4.4 WUR Wake Up Operation

A WUR STA may indicate in the WUR mode request its recommendation on which data rate (LDR or HDR) to use for WUR Wake-up frames if such a rate is supported by the STA as declared in the WUR capabilities. This data rate applies not only to WUR Wake-up frames but also to other frames that the AP might send to this STA (e.g. WUR Short Wake-up frames, keep-alive frames, etc., including group-addressed frames).

Table 6.1 provides a high-level summary on the WUR wake-up operation in three different delivery contexts, including the configurations of the relevant WUR frames and behaviors of the WUR AP and the WUR STA. In this table, Bufferable unit (BU) is used as a shorthand for data or management frames buffered at the AP. If a WUR Wake-up frame contains a Frame Body (FB) field, it is referred to as a VL WUR Wake-up frame (as the Frame Body field is the only field with variable length in a WUR frame). Otherwise, it is referred to as an FL WUR Wake-up frame. As a WUR Short Wake-up frame described later in Section 6.4.5 also has a fixed-length, it is classified as an FL WUR frame in the table as well.

6.4.4.1 Individual DL BU Delivery Context

The first half of this section describes how a WUR AP notifies a WUR STA on unicast BUs. In the second half of this section, corresponding WUR STA behaviors are described.

The WUR AP may transmit a WUR Wake-up frame to a WUR STA to indicate that individually addressed DL BUs are available for the WUR STA. The WUR Wake-up frame includes the following:

- The ID field containing the WUR ID of the WUR STA,
- The ID field contains a WUR group ID that identifies a group of STAs, which include this WUR STA.
- The ID field contains a group ID that identifies a group of STAs including this STA, and one of the identifiers in the FB identifies this STA.

Table 6.1 Summary of MR delivery contexts.

Delivery context	Configurations of the WUR Wake-up frame	Summary
Individual DL BU	**FL WUR Wake Up (unicast or groupcast):** • ID = WUR ID or WUR group ID **VL WUR Wake Up (groupcast)** • ID = WUR group ID • List of WUR IDs in Frame Body	**AP:** Reports that individually addressed DL BUs are available for delivery at the MR for the addressed WUR STA. **STA:** Expected to wake the MR and possibly poll the WUR AP via the MR so that the STA can receive the available DL BUs as soon as possible (e.g. immediately after sending a PS-Poll frame to the AP).
Group addressed DL BU	**FL WUR Wake Up (broadcast):** • ID = transmitter ID or non-transmitter ID **VL WUR Wake Up (not applicable)**	**AP:** Reports that group-addressed DL BUs are available for delivery at the MR for all STAs that are members of the BSS identified in the ID field. **STA:** Expected to wake the MR right before the time the AP is expected to schedule the delivery of the group-addressed DL BUs (e.g. immediately after receiving the DTIM Beacon from the WUR AP) so that the WUR STA can receive group addressed DL BUs of interest.
Critical BSS update	**FL WUR Wake Up (broadcast):** • ID = transmitter ID **VL WUR Wake Up (not applicable)**	**AP:** Reports that critical BSS updates are available at the MR for all WUR STAs that are members of the BSS identified in the ID field or that are members of a BSS that belongs to the same multiple BSSID set as the BSS identified in the ID field. **STA:** Expected to wake the MR right before the time the AP is expected to schedule for transmitting the next Beacon frame so that the WUR STA can receive the Beacon frame.

A WUR AP that transmits a Wake-up frame that indicates the availability of individually addressed BUs follows baseline operation (such as AM mode, PS mode, and TWT) to deliver individually addressed BU(s) to the WUR STA and follow the timing information (e.g. the next service period) that is agreed with the PS operation of MR. Note that the WUR AP can transmit multiple WUR Wake-up frames to increase the likelihood of successful reception.

When the WUR AP schedules for transmitting an MR PPDU (not limited to individually addressed BU(s) to a WUR STA), the WUR AP ensures that either of the following is satisfied:

1) Transition delay indicated by the WUR STA in the WUR capabilities, following the most recent transmitted WUR frame intended to the WUR STA has expired.
2) WUR STA has indicated that it is in the awake state by transmitting a frame to the WUR AP.

There are various optimizations defined for a WUR AP. For example, a WUR AP that generates a VL WUR Wake-up frame orders the WUR STA Info fields in increasing order of AID value. This allows a STA to stop processing the frame once the STA locates a User info field that contains the WUR ID of the STA or a WUR ID that is greater than that of the STA. The WUR AP that sends a WUR Wake-up frame to the WUR STA(s) may send a frame (for example, a Trigger frame) to solicit response frames from one or more WUR non-AP STAs that support the reception of the frame, to improve efficiency of the retrieval of the BUs on the MR. For another example, if the WUR STA and the WUR AP support traffic filtering service (TFS), a procedure to allow more power saving by filtering out unwanted packets, then the WUR AP may use the traffic filter established with the WUR STA to control the generation of WUR Wake-up frames.

After a WUR AP sends a WUR Wake-up frame or a WUR Short Wake-up frame (see Section 6.4.5) with the ID field equal to a WUR ID that identifies a WUR STA, the WUR AP waits for a timeout interval that is larger than the transition delay indicated by the WUR STA in the WUR Capabilities elements:

- If the WUR AP receives any transmission from the WUR non-AP STA within the timeout interval, then the WUR Short Wake-up frame or the WUR Wake-up frame transmission is successful.
- Otherwise, the WUR Short Wake-up frame or the WUR Wake-up frame transmission fails, and the WUR AP may retransmit the WUR (Short) Wake-up frame to the WUR non-AP STA.

The methods by which a WUR AP determines the exact value of the timeout interval and the number of retries after the transmission of individually addressed WUR Wake-up frame fails are implementation specific.

Behavior of a WUR STA is straightforward after it receives a WUR Short Wake-up frame or a WUR Wake-up frame addressed to it with an indication of individually addressed BU(s). The WUR STA simply wakes up its MR to retrieve individually addressed BU(s) based on the procedures of the MR defined in 802.11. While the MR is awake, the WUR STA may put its WUR into the doze state to further reduce the power consumption of the WUR.

6.4.4.2 Group Addressed DL BU Delivery Context

In Section 6.4.1, we described how an AP indicates the presence of group addressed DL BUs and how these BUs are scheduled for delivery via the MR after predetermined DTIM Beacon frames. In addition, we described how a STA in PS mode wakes to receive a DTIM Beacon frame to determine whether group addressed DL BUs that are of interest to the STA are scheduled for delivery, and when this is the case how the STA receives these group addressed DL BUs.

WUR STAs that are in WUR mode, however, are not expected to wake their MR to receive DTIM Beacon frames, and hence they would miss any indication that group addressed DL BUs are scheduled for delivery at the MR. To overcome this issue, the WUR AP is expected to transmit broadcast WUR Wake-up frames to WUR STAs so that they can wake their MR after receiving the broadcast WUR Wake-up frame and prepare for receiving the group addressed DL BUs that are pending at the WUR AP. The context under which these broadcast WUR Wake-up frames are sent is defined as group-addressed DL BU delivery context. Only broadcast FL WUR Wake-up frames can be sent in this context (see Table 6.1).

A WUR AP transmits an FL WUR Wake-up frame with the Group Addressed BU subfield is set to 1 in the Frame Control field to indicate that the WUR frame is sent in the *group addressed DL BU delivery context*. Additionally, the WUR AP sets the ID field of the WUR frame to the transmitter ID if the group addressed DL BU delivery is pending for WUR STAs associated with the transmitted BSSID and sets the ID field of the WUR frame to a non-transmitter ID if the group addressed DL BUs delivery is pending for WUR STAs associated with the non-transmitted BSSID corresponding to that non-transmitter ID. Note that if a WUR AP has group addressed DL BUs available for multiple BSSIDs of the multiple BSSID set then the WUR AP needs to transmit multiple broadcast WUR Wake-up frames, at least one such WUR frame for each of the BSSIDs for which the WUR AP has pending group addressed DL BUs.

Since the reception of group addressed DL BUs is of interest to all the WUR STAs that are members of a given BSS, the WUR AP should attempt to increase the likelihood that all these WUR STAs successfully receive the broadcast WUR Wake-up frame so that they can wake their MR in time for receiving the pending group addressed DL BU frames. To achieve this, the WUR AP may schedule for transmitting multiple broadcast WUR Wake-up frames (benefiting from the obtained redundancy), and possibly distributing these transmissions over multiple WUR DC service periods during which WUR STAs that are in WUR mode are expected to be in WUR awake state (since not all WUR STAs are awake during the same WUR DC SPs). Additionally, the AP should use LDR for these WUR Wake-up frames as opposed to HDR to increase robustness and possibly protect the TXOP during which these WUR frames are sent with NAV-setting frames such as a CTS frame proceeding a WUR transmission.

Now, since WUR STAs need some time to transition their MR from doze state to awake state following the successful reception of a WUR Wake-up frame, i.e. the transition delay, the AP is required to separate the scheduled start time of group addressed DL BUs delivery from the end time of the most recently transmitted WUR Wake-up frame by an interval of time that is at least equal to the maximum transition delay. The maximum transition delay is the highest value among all the transition delays that WUR STAs that are in WUR mode have reported to the AP in their respective WUR Capabilities elements. The maximum transition delay guarantees that even the slowest transitioning STA that happened to receive the last transmitted WUR Wake-up frame has sufficient time to transition the MR from doze state to awake state and be able to receive the pending group addressed DL BUs.

After the maximum transition delay has passed, the WUR AP can then schedule for delivery group addressed DL BU delivery, following the MR protocols that are set in place (see Section 6.4.1).

This enables WUR STAs to wake their MR only for receiving group addressed DL BUs while avoiding the unnecessary periodic wake-ups that need to occur, for example every DTIM, to read DTIM Beacon frames that would contain the indication of group addressed DL BU availability, leading to reduced power consumptions.

Figure 6.7 illustrates an example where a WUR AP wakes up a WUR STA using a WUR Wake-up frame via the WUR to retrieve group-addressed data frames on the MR. Here the WUR STA duty-cycles its WUR radio to detect any BUs for the MR. As the figure shows, the MR of the WUR STA remains in the doze state during the first TIM Beacon frame on the MR, assuming the WUR has not received any WUR Wake-up frame so far. Suppose some group-addressed data frames arrive after the first TIM Beacon frame. Then the WUR AP sends a WUR Wake-up frame via the WUR during the next WUR DC SP X, which wakes up the WUR STA on the MR to receive the frames on the MR. As the WUR STA is synchronized with the WUR AP, the WUR STA does not have to wake up the MR until the next DTIM Beacon frame, as the WUR STA knows that the group addressed data frames are expected to be transmitted only after the DTIM Beacon frame. After receiving all of the data frames, the WUR STA transition back to the doze state on

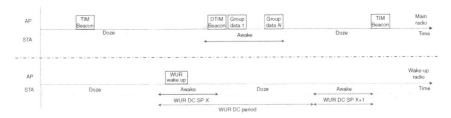

Figure 6.7 An example for the group addressed delivery context.

the MR to save power. In this way, the WUR STA can minimize the time to keep the MR in the awake state for group addressed frames.

6.4.4.3 Critical BSS Update Delivery Context

As discussed in Section 6.4.1, an AP generates Beacon frames in the primary channel of the MR BSS. In addition to the TIM and DTIM information that was discussed previously, these Beacon frames contain a plurality of elements that contain essential information, such as operating parameters, which are needed for the correct functionality of all STAs operating in the BSS. The AP may sporadically update this information, and when such updates occur the AP advertises the updated information in subsequent transmitted Beacon frames.

Since legacy 802.11 STAs are required to wake up periodically to read the Beacon frames, they come to learn of these updates at some point and as such update their operating parameters accordingly. In contrast, WUR STAs are not required to wake up periodically to read the Beacon frames via the MR to maximize their doze times, which in turn leads to a situation where the AP might change some critical parameters of the BSS operation, and WUR STAs might not learn about these changes until it is too late and hence jeopardize the STA's connectivity and/or functionality. In order to avoid this situation, the AP is required to notify the WUR STAs that a critical update has occurred. The AP does so by advertising a BSS parameter update counter to WUR STAs.

By default, the WUR AP provides the most recent value of the BSS parameter update counter in the WUR Operation element included in WUR mode responses sent to the STA, in Beacon frames transmitted via the MR and in broadcast WUR Wake-up frames sent via the WUR. The BSS parameter update counter is provided in the Counter subfield of the WUR Operation Parameters field of the WUR Operation element (see Section 7.2.3) and in the Counter subfield of the TD Control field of the broadcast WUR Wake-up frame (see Section 7.4.2).

This way the WUR AP provides the STA with several means to update its local BSS parameter update counter to the most up-to-date value that the AP is using. When a critical update occurs at the MR the WUR AP increases the value of the BSS Parameter Update Counter and starts advertising this new value in all subsequently transmitted Beacon frames, WUR mode responses and finally in broadcast WUR Wake-up frames. As a consequence, a WUR STA that receives either a WUR response (via the MR) or a broadcast WUR Wake-up frame (via the WUR) whose BSS parameter update counter value is different from the one that the WUR STA is currently storing understands that it needs to wake its MR to obtain critical updates. The STA can obtain the critical updates in a passive manner, by waking up at the next TBTT to receive a Beacon frame sent by the WUR AP, or in an active manner, by probing the WUR AP to generate a Probe Response frame, which will contain the updated BSS parameters. In addition, the STA is required to update its internal BSS parameter update counter to the value in the received frame so that the WUR STA avoids unnecessary wake-ups due to subsequent frames sent by the WUR AP with the same value of the BSS parameter update counter.

The WUR AP notifies the WUR STAs that a critical update is available when any of the following events occurs:

- **Modification of the DSSS parameter set:** This element contains the current channel of the MR BSS, which is the channel where the WUR AP transmits the Beacon frames for the BSS on the MR. A change in the current channel of the MR BSS implies that the WUR STA will not be able to receive any Beacon frames via the MR since the AP has already moved the MR BSS to a new channel and hence suspended transmitting Beacon frames in the current channel. Thus, the WUR STA may need to scan all the available channels to find the new channel where the MR BSS has switched unless the AP sends WUR Discovery frames (see Section 6.2).
- **Modification of the EDCA parameter set:** This element contains enhanced distributed channel access (EDCA) parameters that govern the medium access policies of all QoS STAs operating in the MR BSS. The AP may announce changes in the EDCA parameters of the BSS for various reasons, such as for example the addition of new STAs in the BSS, the generation of new traffic by existing STAs (in the BSS or in neighboring BSSs), or the changes in the offered load.
- **Modification of the HT operation**: This element contains parameters that control the operation of high throughput (HT) STAs that operate in the MR BSS. In particular, it contains the primary channel (which can be located in the 2.4 GHz band or in the 5 GHz band), the BSS bandwidth (which can be either 20 or 40 MHz), the secondary channel if the BSS bandwidth is 40 MHz, and other functionalities that increase protection and coexistence with other neighboring STAs as well as the HT rates that need to be supported when operating in this BSS.
- **Modification of the VHT operation:** This element contains parameters that control the operation of very high throughput (VHT) STAs that operate in the MR BSS. In particular, it contains the BSS bandwidth (which can be 20, 40, 80, 160, or 80 + 80 MHz), the locations of the channel frequency indexes for different VHT BSS operational configurations, and the VHT rates that need to be supported when operating in this BSS.
- **Modification of the HE operation:** This element contains parameters that control the operation of high efficiency (HE) STAs that operate in the MR BSS. Among these parameters the STA can find the BSS bandwidth, and the locations of the channel frequency indexes for different HE BSS operational

configurations when the BSS is located in the 5 GHz band and the AP is not advertising the VHT Operation element and when the BSS is located in the 6 GHz band. In addition, the STA can find information regarding the BSS color in use by the BSS, the HE rates that need to be supported by the STAs when operating in this BSS, etc.

- **Inclusion of a channel switch announcement:** This element announces that the MR BSS is switching from the current channel to a new channel and contains the estimated time when the channel will occur. This solves the issues arising from the modification of the DSSS parameter set since the STA is expected to have enough time to wake the MR and receive a Beacon frame in the current channel, which indicates the new channel and when the switching to this channel will occur.
- **Inclusion of an extended channel switch announcement:** This element provides an extended functionality compared to the Channel Switch Announcement because it enables signaling a channel switch between different bands as well, e.g. from 2.4 to the 5 GHz band.
- **Inclusion of a wide bandwidth channel switch:** This element is included along with an (extended) channel switch announcement to provide additional operating information for the MR BSS in the new channel, such as the new BSS bandwidth and the new channel center frequencies.
- **Inclusion of a channel switch wrapper:** This element acts as a wrapper for a plurality of sub-elements, such as New Country, Wide Bandwidth Channel Switch, and New Transmit Power Envelope sub-elements, and is included along with an extended channel switch announcement. The New Country sub-element is generally present when the BSS is transitioning to a new country, in which case the AP provides the necessary configurations, in terms of operating classes tables and sets, that the STAs need to follow in the new channel. The Wide Bandwidth Channel Switch sub-element provides new BSS bandwidth, and channel center frequencies, while the New Transmit Power Envelope sub-elements provide maximum transmit power requirements that need to be followed within the BSS in the new channel

While the above is a list of information elements supporting MR functionalities, whose changes are required to be classified as critical updates, the AP may classify other updates as critical ones as well, e.g. modification of RSNE, and so on. In other words, the critical update in IEEE 802.11 is extensible. On the MR, the TIM Broadcast feature defines the list of elements whose changes are treated as a critical update, and the same list has been adopted by the critical update procedure in 802.11ba. If the TIM Broadcast feature on the MR includes a new element to the list for a critical update in a future 802.11 standard, changes to the new element should also be advertised as a critical update by a broadcast WUR Wake-up frame.

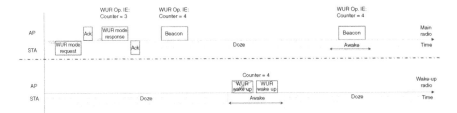

Figure 6.8 An example for the critical BSS update delivery context.

Figure 6.8 illustrates an example where a WUR AP wakes up a WUR STA using WUR Wake-up frames via the WUR to obtain a recent critical update indicated in Beacon frames. In the beginning, the WUR STA obtained the latest BSS parameter update counter of three using the WUR Mode Request/Response frames on the MR. Suppose one of the HE operation parameters has changed, which increases the counter to four in the next Beacon frame. As the MR of the WUR STA remains in the doze state to save power during the Beacon frame, the WUR STA fails to learn the critical update from this Beacon frame. To help the WUR STA learn the update as soon as possible, the WUR AP transmits two duplicate WUR Wake-up frames when the WUR STA transitions to the awake state in the next WUR DC SP. The duplicate WUR Wake-up frames are shown in this example to illustrate how to increase the likelihood of successful wake-up even if the first WUR Wake-up frame fails to be received by the WUR STA due to interference. As the WUR STA is synchronized with the WUR AP, the WUR STA does not have to wake up the MR until the next Beacon frame, through which the WUR STA obtains the latest HE operation parameters. In this way, the WUR STA can minimize the time to keep the MR in the awake state for critical BSS update on the MR.

6.4.5 Use of WUR Short Wake-up Frames

In the previous section, we discussed how the WUR AP causes the MR of WUR STA(s) to wake up by transmitting WUR Wake-up frames. In this section, we continue by describing the use of WUR Short Wake-up frames instead of WUR Wake-up frames in a particular context. WUR Short Wake-up frames have smaller size compared to WUR Wake-up frames (see Section 7.4.5), which translates to shorter PPDU transmit times, e.g. 0.668 ms as opposed to 0.924 ms when LDR is used. Shorter PPDU transmit times are particularly beneficial when the channel utilization is high.

A WUR AP may send WUR Short Wake-up frames to a WUR STA if both the AP and the STA have declared their support in the WUR Short Wake-up frame Support subfield of their respective WUR Capabilities elements. WUR Short

Wake-up frames are sent instead of WUR Wake-up frames only when the WUR AP intends to wake the MR of a WUR STA for delivering pending DL BUs intended for the WUR STA, which is described in individual DL BU delivery context in Section 6.4.4. WUR Short Wake-up frames cannot be used in the remaining contexts described in Section 6.4.4, such as group DL BU delivery, critical BSS update delivery, etc., since WUR Short Wake-up frames do not carry the required signaling for these additional functionalities. Following the reception of the WUR Short Wake-up frame, the WUR STA is expected to be in the awake state at the earliest SP as described under the individual DL BU delivery context.

In order to provide added security, the WUR AP that has established a secure association with the WUR STA is required to provide to the WUR STA a random WUR ID every time that the WUR STA wakes up as a result of the reception of a WUR Short Wake-up frame and successfully exchanges frames with the WUR AP via the MR. The WUR ID is provided to the WUR STA in an unsolicited WUR Mode Response frame that is sent under the context of a WUR mode update (see Section 6.4.2). The WUR Mode Response frame should be encrypted (i.e. protected) to ensure that no eavesdroppers overhear the new WUR ID. The WUR AP should not retransmit a WUR Short Wake-up frame (i.e. with the same WUR ID) to the WUR STA, but may retransmit a WUR Wake-up frame instead.

On the other side, the WUR STA shall only accept as valid one received WUR Short Wake-up frame that matches the locally stored WUR ID. I.e. the STA shall ignore all subsequently received WUR Short Wake-up frames with matching ID until a new WUR ID has been configured at the WUR STA by the WUR AP.

Now, there are cases where the WUR STA wakes its MR without having received any WUR Wake-up frame. For example, the STA has uplink frames to send to the AP, and so on. When this happens, the WUR STA may transmit to the AP a WUR Wake-up Indication frame as the first frame. The reception of a WUR Wake-up Indication frame by the AP is an indication that the WUR STA has performed an unsolicited wake-up and as such the WUR STA has not used the previously configured WUR ID. Hence, there is no need for the WUR AP to randomly create a new WUR ID to provide to the WUR STA in the case of an unsolicited wake-up of the MR. If the WUR STA does not send WUR Wake-up Indication frame that indicates an unsolicited wake-up, then this could mean that the WUR STA has woken up due to a received WUR Short Wake-up frame. Now, if the WUR AP had not sent a WUR Short Wake-up frame, then it may be that there is a "man in the middle attack." In other words an attacker is performing a brute force attack, or is eavesdropping the connection between the WUR AP and the WUR STA, obtaining the WUR IDs and using them to send WUR Short Wake-up frames to the WUR STA. This, in turn, causes unnecessary wake-ups of the MR of the STA, increasing the power consumption of the STA, which can potentially lead to battery drainage. In order to mitigate this type of attack, the AP should invoke a timeout before configuring a new WUR ID for the STA, and in the meantime revert back to using WUR Wake-up frames, in particular protected WUR Wake-up frames, which are described in Section 6.6. The AP should double this interval every time that the STA sends frames via the MR but does not send a WUR Wake-up Indication frame indicating unsolicited wake-up. This is because it is likely that the attack is still persisting. When this occurs, the STA might not respond to WUR Short Wake-up frames until a new WUR ID is configured by the associated AP.

6.4.6 Keep Alive Frames

In the previous sections, we described protocols that serve different purposes, such as discovery via WUR Discovery frames, synchronization via WUR Beacon frames, and MR wake-up via WUR (Short) Wake-up frames. From a scheduling perspective, these protocols are divided into two broad categories:

1) **AP centric:** The schedule for generating these WUR frames is determined by the WUR AP. The generation of WUR Beacon and WUR Discovery frames falls in this class. These are broadcast WUR frames and as such they are intended for all WUR STAs that are expected to interact with the WUR AP one way or another. The reception of these frames is an indication to the WUR STA that it is still connected with the WUR AP (i.e. the WUR connection is "alive").

2) **STA centric:** The schedule for generating the WUR frames is determined by traffic and DC configuration of each WUR STA. The generation of WUR Wake-up and Short Wake-up frames falls in this class.

It is worth noting that the schedule for generating WUR frames that fall in the keep-alive category falls in the WUR AP-centric protocols, which means that the AP determines the periodicity itself without any input by any WUR STAs that are associated with the WUR AP. In some cases, however, it is beneficial for a WUR STA to provide input as to how often these keep-alive frames are generated.

A WUR STA negotiates the use of keep-alive WUR frames, along with the other WUR parameters that were described in Section 6.4.2, during WUR mode setup. In particular, the WUR STA requests the generation of keep-alive WUR frames in the Requested Keep Alive WUR Frame subfield of the WUR Mode element included in the WUR Mode Request frame that it transmits to the WUR AP. The WUR AP then indicates in the WUR Mode Response frame whether the request is accepted or denied. The WUR AP can indicate that the request is denied due to impossibility of generating keep-alive frames.

If the request is accepted, then the WUR AP shall schedule for transmitting a keep-alive WUR frame during each negotiated WUR DC service period. The following frames can be used to provide keep-alive functionality (i.e. as keep-alive WUR frames):

- **WUR Beacon frame:** Expected to be the default keep-alive frame that are sent within WUR DC service periods during which the WUR AP does not intend to cause the MR of the WUR STA to wake up. Another benefit of using the WUR Beacon frames is that they help the WUR STA in synchronizing the TSF timer during every negotiated WUR DC service period without the need of waking every TWBTT. And since WUR Beacon frames are broadcast these functionalities are also provided to other WUR STAs that are associated to the WUR AP and that happen to be awake and that receive these "additional" WUR Beacon frames generated by the WUR AP.
- **WUR Wake-up or Short Wake-up frame:** Expected to be sent instead of WUR Beacon frames only within specific WUR DC service periods. This is because a side effect of the reception of these frames is that they cause the MR of the WUR STA to wake up and interact with the AP by exchanging frames via the MR as described in Section 6.4.4, leading to increased power consumption at the WUR STA due to MR utilization. There are also some benefits with the WUR Wake-up or Short Wake-up frames in comparison with WUR Beacon frames. They can allow a WUR STA to go back to doze state earlier and can be transmitted on a channel that is not a WUR primary channel, which is needed if a WUR STA uses FDMA described in the Section 6.5.

6.5 Frequency Division Multiple Access

In Section 6.4, we described several WUR protocols according to which the WUR AP sends a WUR frame to provide a certain functionality or service (e.g. wake up) to a specific WUR STA or a group of WUR STAs. The WUR frame in these cases is carried in a WUR Basic PPDU that has a bandwidth of 20 MHz and is transmitted by the WUR AP in the WUR primary channel. The WUR Basic PPDU, depending on the data rate being used and the length of the carried WUR frame, may require a fair amount of airtime to be transmitted, i.e. the transmit duration can be up to ~3 ms (e.g. this maximum limit is reached in the case of a VL WUR Wake-up frame addressed to a list of 8 WUR STAs and sent with LDR).

To improve the scalability of WUR operation, WUR FDMA operation is defined to enable the transmission of multiple WUR Wake-up frames multiplexed in frequency. This is motivated by the fact that a WUR AP can operate on a bandwidth that is multiples of 20 MHz. For example, many APs operate with an 80 MHz bandwidth. This means that up to 4 WUR Basic PPDUs may be transmitted in parallel within a single TXOP instead of sending them one after another using 4 TXOPs. Such improvement is especially helpful for a WUR AP with a wide operating bandwidth and a large number of associated WUR STAs such as IoT devices. This is an optional feature and a WUR AP or WUR STA

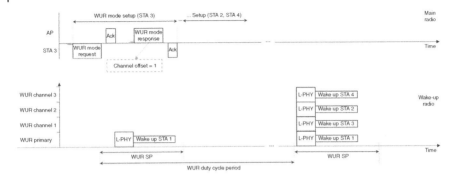

Figure 6.9 Operation with WUR FDMA PPDUs.

advertises its support of the WUR FDMA operation in the WUR FDMA Support subfield of the WUR Capabilities element.

WUR FDMA is negotiated between a WUR AP and multiple WUR STAs, as illustrated in Figure 6.9. The WUR AP enables WUR FDMA operation with multiple WUR STAs during WUR mode setup via the MR. FDMA operation is established with a WUR STA after a successful WUR mode setup. In this example, the AP allocates 4 WUR STAs to 4 orthogonal WUR channels, allowing up to 4 WUR STAs to be woken up in parallel within a short WUR SP using a WUR FDMA PPDU (see Section 4.8). The WUR FDMA PPDU may contain any WUR frame types including WUR Beacon frame, WUR Wake-up frame, WUR Short Wake-up frame, WUR Discovery frame, and WUR Vendor Specific frame. If a keep-alive WUR Beacon frame is carried in the WUR FDMA PPDU, the AP ensures that the frame is transmitted within the WUR primary channel.

The WUR Model element (see Section 7.2.3) provides flexibilities for the WUR AP and WUR STA to negotiate how to assign a WUR channel to the WUR STA. The WUR AP may set the WUR Channel Offset subfield in the WUR Mode element to a nonzero value if the WUR STA supports WUR FDMA, indicating a WUR channel that is above or below the WUR primary channel (see Table 7.5). Otherwise, the value shall be set to 0, indicating the WUR primary channel.

If the WUR AP supports WUR FDMA, the WUR STA has an option to recommend the AP to assign a particular WUR channel to the WUR STA in the WUR Parameters field of the WUR Model element. Such recommendation is beneficial if there is some channel condition such as interference only visible at the WUR STA but not at the WUR AP. During the WUR mode setup phase the WUR STA can additionally negotiate whether HDR or LDR can be used for the WUR frames to be transmitted to the WUR STA. After receiving the recommendation from the WUR STA, the WUR AP has the final say on which channel is to be assigned for the WUR STA.

There are a few rules on how the AP schedules WUR FDMA transmissions. In general, the WUR AP will ensure that the WUR SP does not overlap with TWBTT at which the WUR AP has scheduled transmission of WUR Beacon frames. This is because the WUR STA is expected to wake and receive the WUR Beacon frame for timing and synchronization purposes and as such is not able to be in both subchannels at the same time, essentially leading to the WUR STA losing at least one of these frames.

The WUR AP can allocate the WUR frames addressed to different WUR STAs in different subchannels. When the WUR AP schedules WUR Wake-up frames for the multiple WUR STAs, the AP transmits a WUR FDMA PPDU as shown in Figure 6.9. If the frames have different lengths, the WUR AP ensures that the WUR FDMA PPDU starts and ends at the same time in the different subchannels by padding.

The WUR AP can also transmit a punctured WUR FDMA PPDU to handle dynamics in interference or in the number of pending frames. For example, if the WUR AP detects that all WUR Channels are idle except for WUR Channel 1 during an interval of PIFS immediately preceding the start of the WUR FDMA PPDU transmission, the AP may transmit a punctured WUR FDMA PPDU occupying all the WUR channels except for WUR Channel 1. Essentially WUR STAs 1, 2, and 4 can still be woken up by this single PPDU. The same punctured FDMA PPDU may also be used if the WUR AP does not have a pending frame intended for the STA 3.

It is worth noting that a WUR FDMA PPDU always occupies the WUR primary channel, to help the WUR AP keep control of the WUR primary channel during the PPDU transmission. If the WUR AP does not have a pending WUR frame intended for WUR non-AP STAs on the WUR primary channel, then the WUR AP shall transmit a WUR frame on the WUR primary channel, which may be any WUR frame that does not cause a WUR non-AP STA to wake up on the WUR primary channel.

6.6   Protected Wake-up Frames

A WUR AP uses protected WUR frames to improve efficiency and security of transmissions over the WUR. Wireless security is a big topic by itself, so we mainly try to provide a high-level overview on the unique security designs in 802.11ba in this section.

The protected WUR frames help a WUR STA in many ways. For a battery-power device such as a security camera, they can prevent an attacker from waking up the MR frequently, which is one possible attack that drains the power and disable the device in a short time.

The protected WUR frames improve security by enabling integrity check. As the size of the WUR frames needs to be minimized, the WRU AP sets the last 2 octets of a WUR frame, named the frame check sequence (FCS) field, to a message integrity check (MIC) value for integrity check. If an attacker attempts to create a fake WUR frame or replay a previously overheard legitimate WUR frame, a receiving WUR STA is able to discard the frame based on its MIC value. If integrity check is not needed, the WUR AP sets FCS field to cyclic redundancy check (CRC) for transmission error detection. At high level, the protected WUR frames utilizes the MIC value in the FCS field for both integrity check and transmission error detection. This design choice helps to minimize the sizes of protected WUR frames while protecting a STA from attacks such as replay attack.

To make it hard for an attacker to create a fake WUR frame, a WUR AP and its associated WUR STAs share some secrete keys, such as the wake-up radio temporal key (WTK) for frames towards a single WUR STA or the wake-up radio integrity group temporal key (WIGTK) for frames towards a group of WUR STAs. To prevent an attacker from sniffing a legitimate protected WUR frame over the air and replaying it later, the WUR AP and WUR STAs make sure that a new protected WUR frame, even identical to a previously transmitted protected WUR frame except for the FCS field, should have a different MIC value. As a replayed frame contains a stale MIC value, the frame will be discarded. The uniqueness of the MIC values is achieved by generating the MIC values based on an always-changing packet number (PN) maintained by the WUR AP and WUR STAs.

6.7 Conclusion

In this chapter we have focused on the main MAC functionalities of 802.11ba. However, 802.11ba has more to offer. For example, 802.11ba has introduced a WUR Vendor Specific frame that can contain information beyond what has been defined in 802.11ba. This frame allows vendors to provide supplementary information that may enhance WUR functionalities and/or address additional applications or use cases that were not within the scope of IEEE 802.11ba at the time of standardization.

In conclusion, this chapter describes how WUR is integrated to and improves network discovery, connectivity, time synchronization, and power management in an 802.11 network. This chapter also covered FDMA and protected WUR frames that improve efficiency and security for transmissions over the WUR. The following chapter will cover the detailed frame formats that support the MAC functionalities discussed in this chapter.

7

Medium Access Control Frame Design

7.1 Introduction

The goal of this chapter is to help the readers, even those unfamiliar with the IEEE 802.11 standard, to quickly grasp the idea of how information that is relevant to the functionalities discussed in previous chapters is signaled in IEEE 802.11 frames.

We will first go over the design details for the information elements, then main radio (MR) medium access control (MAC) frames and wake-up radio (WUR) MAC frames. Information elements are covered in Section 7.2, MR MAC frames are covered in Section 7.3, and we conclude with WUR MAC frames in Section 7.4.

7.2 Information Elements

7.2.1 General

Information elements are small containers that carry management information. This information is shared by including one or more elements in Management frames that are exchanged between devices (see Section 7.3). Each element is identified by a numerical label and has a length that depends on the amount of information carried by the element.

Since its inception the IEEE 802.11 standard has defined a variety of information elements, each of which carrying information for specific functionalities and/or amendments. The number of elements is expected to grow further in the future as more functionalities are defined by the standard. In the early days, each element was identified by an element ID of 8 bits and as such up to 256 could be defined. However, when a specific amendment (namely IEEE 802.11ai) was being proposed, the IEEE 802.11 working group realized that they were running out of available element IDs.

IEEE 802.11ba: Ultra-Low Power Wake-up Radio Standard, First Edition.
Steve Shellhammer, Alfred Asterjadhi, and Yanjun Sun.
© 2023 The Institute of Electrical and Electronics Engineers, Inc.
Published 2023 by John Wiley & Sons, Inc.

To solve this issue a second numerical label was introduced, which extends the element's identifiers range, namely the Element ID Extension field.

Elements have a common general format that is shown in Figure 7.1.

Each element is identified by the value of the Element ID field, and if present, the Element ID Extension field. A specific value of the Element ID field, namely the value 255, is used as an escape value to indicate that the Element ID Extension field is present, in which case the element will be identified by the combination of the Element ID field (which has the value 255), and the Element ID Extension field.

The Length field indicates the number of bytes in the element. More precisely the Length field indicates the length of the Information field if the Element ID Extension field is not present, and it indicates the length of the Information field plus one if the Element ID Extension field is present. The Information field carries information specific to the element.

The structure of the information element is designed in such a way that it ensures future expansibility, i.e. future amendments can include additional information in the same element while ensuring backwards compatibility. However, for simplicity, during the early days of standards development, and more rarely later on, some elements were defined as non-extensible, while more recently most of the elements are defined as extensible. An element that is extensible may change its size in the future due to the addition of more fields, which can be parsed by newer devices but not by older devices. The older devices in this case can parse only the information, up to the length they have implemented to parse, and ignore the rest of the element.

In the following sections, we will cover the details of several elements that are defined in the standard in support of the functionalities discussed in the previous chapters. These elements are used for exchanging management information related to MR functionalities (covered in Section 7.2.2) and information related to WUR functionalities (covered in Section 7.2.3).

7.2.2 Elements Supporting MR Functionalities

This section describes elements that are used for exchanging information related to key MR functionalities, such as operating channel, enhanced distributed channel access (EDCA) parameters, etc. When there is a change in any of these parameters, the critical BSS update procedure described in Section 6.4.4 helps a WUR Station

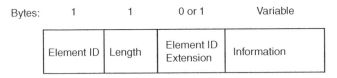

Figure 7.1 Element format.

(STA) determine that a change has occurred in a timely manner. The goal of this section is to provide a general description of the key fields contained in these elements so that the reader can have a high-level understanding of these parameters and use this section as a quick reference to understand previous chapters without reading the detailed IEEE 802.11 standard.

7.2.2.1 DSSS Parameter Set Element

The DSSS Parameter Set element advertises the channel in the 2.4 GHz band that is currently in use by the BSS. The DSSS Parameter Set element is decodable by all IEEE 802.11 devices that support direct sequence spread spectrum (DSSS) PHY/ MAC functionalities, which are essentially all STAs that support any of the IEEE 802.11b/g/n/ax amendments in the 2.4 GHz band, preserving backward compatibility. A WUR AP indicating during WUR discovery (see Section 6.2.2), that an MR AP is operating in the 2.4 GHz band needs to ensure that the advertised operating channel in the WUR Discovery frame is aligned with the channel being advertised in the DSS Parameter Set element sent by the MR AP.

The format of the DSSS Parameter Set element is shown in Figure 7.2.

The Current Channel field contains the channel number of the current channel of the BSS.

7.2.2.2 EDCA Parameter Set Element

The EDCA Parameter Set element advertises the EDCA parameters that control the wireless medium access when exchanging frames within the BSS that has been established by the AP. These parameters are used by the distributed coordination function (DCF) and the EDCA mechanisms, which were briefly described in Section 2.3.4.

The format of the EDCA Parameter Set element is shown in Figure 7.3.

The QoS Info field sent by an AP contains a series of subfields which are briefly summarized below:

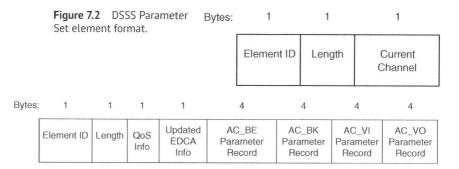

Figure 7.2 DSSS Parameter Set element format.

Figure 7.3 EDCA Parameter Set element format.

- EDCA Parameter Set Update Count subfield contains a counter that is incremented each time any of the announced EDCA Parameter Record fields change.
- Q-Ack subfield is a flag that indicates whether the AP allows associations with STAs advertising support for the Q-Ack functionality, which can be used within a particular channel access protocol, namely hybrid contention channel access (HCCA).
- Queue Request subfield is a flag that indicates whether the AP is capable of processing queue sizes in the QoS Control field of received QoS Data frames.
- TXOP Request subfield is a flag that indicates whether the AP is capable of processing TXOP duration requests in the QoS Control field of received QoS Data frames.

The Updated EDCA Info field contains information that is used by STAs that operate in the sub 1 GHz band, which is out of the scope of this book.

All of the AC_X Parameter Record fields have the same common format, which is shown in Figure 7.4.

The ACI /AIFSN subfield contains the following subfields:

- AIFSN subfield indicates the number of slots, after a short interframe space (SIFS), that an STA needs to defer prior to either invoking the EDCA backoff or starting a transmission. The obtained parameter, arbitration interframe space (AIFS), is obtained as SIFS + AIFSN*Slot.
- ACM subfield indicates if admission control is required prior to transmitting using the access parameters specified for this access category (AC).
- ACI subfield contains the access category index from which one can obtain the AC, namely the value 0 indicates AC_BE (best effort), value 1 indicates AC_BK (background), value 2 indicates AC_VI (video), and value 3 indicates AC_VO (voice).

The ECWmin/ECWmax subfield contains the following subfields:
- ECWmin subfield indicates the value of the minimum contention window, CW_{min}, in an exponential form, i.e. $CW_{min} = 2^{ECWmin} - 1$.

- ECW_{max} subfield indicates the value of the maximum contention window, CW_{max}, in exponential form, i.e. $CW_{max} = 2^{ECWmax} - 1$.

The TXOP Limit field indicates the maximum duration of any obtained TXOP for this particular AC, except for the value 0, which has a special meaning.

Bytes: 1 1 2

| ACI/AIFSN | ECWmin/ECWmax | TXOP Limit |

Figure 7.4 AC_X Parameter Record field format.

The transmit opportunity (TXOP) is an interval of time during which a STA has the right to initiate frame exchanges over the wireless medium without the need of contending again for accessing the medium.

Each TXOP has a starting time and a maximum duration. The starting time is the start time of the slot at which the STA gains access to the medium, after contending for the medium using EDCA, which is governed by AIFS, CWmin, and CWmax, and the maximum TXOP is the maximum duration of the TXOP, governed by the TXOP Limit.

7.2.2.3 Channel Switch Announcement Element

The Channel Switch Announcement element announces that the BSS is switching from the currently advertised channel to a new channel and the estimated time when the channel switch will occur. This element can only advertise a channel switch within a specific band (i.e. the channel switch can occur within a specific band, for example the BSS can switch from channel 1 to channel 5 in the 2.4 GHz band).

The format of the Channel Switch Announcement element is shown in Figure 7.5.

The Channel Switch Mode field indicates whether there are any restrictions on the transmissions in the current channel until the channel switch occurs. The value 0 indicates that there are no such restrictions imposed to the STA receiving this element. On the other hand, the value 1 indicates that the STA receiving this element is not allowed to transmit any additional frames on the current channel until the scheduled channel switch occurs. The STA may transmit frames on the new channel after the channel switch has occurred.

The New Channel Number field indicates the number of the new channel to which the STA is switching.

The Channel Switch Count field is set to the number of TBTTs until the STA switches to the new channel. In particular, the value 1 indicates that the switch will occur at the next TBTT, in which case the Beacon that is scheduled at that TBTT will be sent on the new channel and will advertise information of that channel. The value 0 instead indicates that the switch will occur at any time after the frame containing this element is transmitted.

Bytes:	1	1	1	1	1
	Element ID	Length	Channel Switch Mode	New Channel Number	Channel Switch Count

Figure 7.5 Channel Switch Announcement element format.

7.2.2.4 Extended Channel Switch Announcement Element

The Extended Channel Switch Announcement element announces that the BSS is switching from the currently advertised channel to a new channel, with the new channel being in the same operating class or in a new operating class. This element expands the functionalities provided by the Channel Switch Announcement element, described above, by enabling a channel switch across bands (for example, the BSS can switch from channel 1 in the 2.4 GHz band to channel 36 in the 5 GHz band, where the band is specified by the operating class).

The format of the Extended Channel Switch Announcement element is shown in Figure 7.6.

The Channel Switch Mode, New Channel Number, and the Channel Switch Count fields are similar to their respective fields in the Channel Switch Announcement element, which were described above, except that the new channel number is a channel selected from the STA's new operating class.

The New Operating Class field indicates the new operating class and helps identify the band within which the new channel is located.

7.2.2.5 HT Operation Element

The HT Operation element controls the operation of high throughput (HT) STAs that operate in the BSS, i.e. STAs that support the IEEE 802.11n amendment. Compared to previous amendments, IEEE 802.11n, among other things, added support for operating in the 2.4 and in the 5 GHz band, increased BW support from 20 to 40 MHz, and increased number of spatial streams (NSS) support from 1 to 4. The HT Operation element is decodable by all IEEE 802.11 devices that support the HT PHY/MAC functionalities, which are essentially all HT STAs and all future STAs that support amendments built on top of IEEE 802.11n in the specific band (i.e. IEEE 802.11ax in the 2.4 GHz band and IEEE 802.11ac/ax in the 5 GHz band).

The format of the HT Operation element is shown in Figure 7.7.

The Primary Channel field contains the channel number of the primary channel of the BSS.

The HT Operation Information field contains several subfields related to HT functionalities that are currently in use in the BSS such as:

Bytes:	1	1	1	1	1	1
	Element ID	Length	Channel Switch Mode	New Operating Class	New Channel Number	Channel Switch Count

Figure 7.6 Extended Channel Switch Announcement element format.

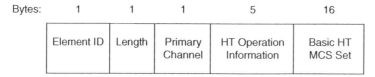

Bytes: 1   1   1   5   16

Element ID	Length	Primary Channel	HT Operation Information	Basic HT MCS Set

Figure 7.7 HT Operation element format.

- Secondary Channel Offset subfield indicates the location of the secondary channel of the BSS with respect to the primary channel, which can be, if present, either above or below the primary channel.
- STA Channel Width subfield indicates the BSS bandwidth for the HT BSS. All transmissions from HT STAs within this BSS have to be less than or equal to the HT BSS bandwidth (20 or 40 MHz).
- HT Protection subfield indicates the protection requirements for HT transmissions: no protection, nonmember protection, 20 MHz protection, and non-HT mixed protection.
- Non-greenfield HT STAs Present subfield indicates whether any HT STAs that are not HT-greenfield capable have joined the BSS, which, if it is the case, would have the HT STAs in the BSS enable additional protection mechanisms for ensuring coexistence with these other STAs.
- OBSS Non-HT STAs Present subfield indicates whether any STAs that are not HT STA capable have joined a neighboring BSS, which, if it is the case, would have the HT STAs in the BSS enable additional protection mechanisms for ensuring coexistence with these other STAs.
- Channel Center Frequency Segment 2 subfield indicates the channel center frequency for a 160 or 80 + 80 MHz BSS BW for a particular class of VHT STAs (see Section 7.2.2.6) and is ignored by HT STAs that are not VHT STAs.
- RIFS Mode, Dual Beacon, Dual CTS Protection, and STBC Beacon subfields are not covered since these functionalities have become obsolete.

The Basic HT MCS Set field indicates which HT-MCS values are to be supported by all HT STAs that operate in the BSS.

7.2.2.6   VHT Operation Element

The VHT Operation element controls the operation of very high throughput (VHT) STAs that operate in the BSS, i.e. STAs that support the IEEE 802.11ac amendment. Because VHT STAs are also HT STAs their operation is controlled by the HT Operation element as well, which was discussed above. Compared to previous amendments, IEEE 802.11ac, among other things, increased BW support from 40 to 160 MHz, increased NSS support from 4 to 8, and added downlink multi-user (MU) multiple input multiple output (MIMO). VHT STAs only operate in the 5 GHz band mainly because BWs wider than 40 MHz cannot be used in the 2.4 GHz band.

The VHT Operation element is decodable by all IEEE 802.11 devices that support VHT PHY/MAC functionalities, which are essentially all VHT STAs and all future STAs that support amendments built on top of IEEE 802.11ac in the 5 GHz band (i.e. IEEE 802.11ax).

The format of the VHT Operation element is shown in Figure 7.8.

The VHT Operation Information field contains several subfields related to VHT functionalities in use in the BSS such as:

- Channel Width subfield, which, together with the STA Channel Width field of the HT Operation element, indicates the BSS bandwidth for the VHT BSS. All transmissions from VHT STAs within this BSS have to be less than or equal to the VHT BSS bandwidth (20, 40, 80, 160, or 80 + 80 MHz).
- Channel Center Frequency Segment 0 indicates a channel center frequency for 20, 40, 80, 160, or 80 + 80 MHz VHT BSS.
- Channel Center Frequency Segment 1 indicates a channel center frequency for 160, or 80 + 80 MHz VHT BSS.

These channel center frequency segments (eventually along with the channel center frequency segment 2, which is included in the HT Operation element) identify the channel frequency indexes for different VHT BSS operational configurations, which depend on the BSS bandwidth, and the location of the primary channels and of the secondary channels.

The Basic VHT-MCS and NSS Set field indicates the VHT-MCSs values for each NSS that are to be supported by all VHT STAs in the BSS.

7.2.2.7 Wide Bandwidth Channel Switch Element

The Wide Bandwidth Channel Switch element announces the BSS bandwidth and the BSS channel center frequencies that the BSS will use once the (extended) channel switching has concluded, i.e. after the BSS has moved to the new channel.

The format of the Wide Bandwidth Channel Switch element is shown in Figure 7.9.

The New Channel Width field indicates the BSS bandwidth (20, 40, 80, 160, or 80 + 80 MHz) that the BSS will have after switching to the new channel.

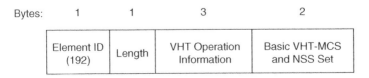

Bytes:	1	1	3	2
	Element ID (192)	Length	VHT Operation Information	Basic VHT-MCS and NSS Set

Figure 7.8 VHT Operation element format.

Figure 7.9 Wide Bandwidth Channel Switch element format.

The New Channel Center Frequency Segment 0 and Channel Center Frequency Segment 1 fields indicate the channel frequency indexes that the BSS will have after switching to the new channel.

7.2.2.8 Channel Switch Wrapper Element

The Channel Switch Wrapper element contains a list of subelements that provide additional configurations that the BSS will use once the extended channel switching has concluded. These configurations include providing new country information (e.g. when the BSS moves from one country to another), BSS bandwidth and channel information in the new channel, and eventually transmit power requirements for the BSS in the new channel and/or new country.

The format of the Channel Switch Wrapper element is shown in Figure 7.10.

The New Country subelement is present when the AP is performing extended channel switching to a new country, or when switching to a new operating classes table/set. The subelement indicates the country where the BSS will be located, the operating class table, and the operating classes that the BSS will be using in the new channel.

The Wide Bandwidth Channel Switch subelement is present if the AP is performing extended channel switching and is also changing the BSS bandwidth and/or the BSS channel center frequencies that the BSS will be using in the new channel.

The New Transmit Power Envelope subelement(s) is/are present if the AP is performing extended channel switching and is also changing the transmit power requirements that the BSS will be following in the new channel. Each New Transmit Power subelement indicates the maximum transmit powers that will be followed within the BSS in the new channel.

7.2.2.9 HE Operation Element

The HE Operation element controls the operation of high efficiency (HE) STAs that operate in the BSS, i.e. STAs that support the IEEE 802.11ax amendment. Because

Bytes:	1	1	Variable	Variable	Variable
	Element ID (196)	Length	New Country Subelement (optional)	Wide Bandwidth Channel Switch Subelement (optional)	New Transmit Power Envelope Subelement(s) (optional)

Figure 7.10 Channel Switch Wrapper element format.

HE STAs are also HT STAs in the 2.4 GHz band and are also VHT STAs in the 5 GHz band, their operation is controlled by the HT Operation element and the VHT Operation element as well. Compared to previous amendments, IEEE 802.11ax, among other things, introduced downlink and uplink orthogonal frequency-division multiple access (OFDMA), uplink (UL) MU MIMO, spatial reuse procedures, target wake time (TWT), and operation in the 6 GHz band. The HE Operation element is decodable by all IEEE 802.11 devices that support HE PHY/MAC functionalities, which are currently all HE STAs and possibly all future STAs that support amendments built on top of IEEE 802.11ax in these bands.

The format of the HE Operation element is shown in Figure 7.11.

The HE Operation Parameters field contains a series of subfield which are described below:

- Default PE subfield indicates the default duration of the packet extension (PE) for HE trigger-based (TB) PPDUs that are sent in response to certain HE MU PPDUs (more specifically one that contains MPDUs with triggered response scheduling (TRS) control fields in them). The PE is essentially PHY padding needed by the AP so that it has enough time to process the received HE TB PPDU and generate the immediate control response that would be required by the STA that is sending the HE TB PPDU.
- TWT Required indicates that the AP requires all associated STAs that support TWT to setup TWT schedules with the AP.
- TXOP Duration RTS Threshold indicates a time threshold beyond which all TXOPs that are to be obtained by associated STAs need to be preceded by an RTS/CTS exchange sequence.
- VHT Operation Information Present subfield indicates whether the VHT Operation Information field is present in the HE Operation element. The VHT Operation Information field is required to be present when included in a frame that is sent in the 5 GHz band that contains the HT Operation element and does not contain the VHT Operation element.
- Co-Hosted BSS subfield indicates whether the AP is a member of a co-hosted BSSID set. A co-hosted BSSID set consists of BSSs that operate on the same channel, are managed from the same physical AP, and require independent handling (i.e. each BSS has its own beacons, time synchronization functions, security, etc.) as opposed to a multiple BSSID set which require unified handling for most functionalities.

Bytes	1	1	1	3	1	2	0 or 3	0 or 1	0 or 5
	Element ID	Length	Element ID Extension	HE Operation Parameters	BSS Color Information	Basic HE-MCS and NSS Set	VHT Operation Information	Max Co-Hosted BSSID Indicator	6 GHz Operation Information

Figure 7.11 HE Operation element format.

- ER SU Disable subfield indicates whether extended range (ER) single user (SU) PPDUs are allowed or not in the BSS.
- 6 GHz Operation Information Present subfield indicates whether the 6 GHz Operation Information field is present in the HE Operation element. The 6 GHz Operation Information field is required to be present when the frame containing the HE Operation element is sent in the 6 GHz band.

The BSS Color Information field contains a series of subfields which are described below:

- BSS Color subfield contains a non-unique identifier for the BSS or for the set of the BSSs that belong to a multiple BSSID set or a co-hosted BSSID set. The BSS color is then included in the PHY header of all HE PPDUs that are generated within the BSS. This identifier helps enabling different functionalities such as, spatial reuse, intra-PPDU power saving, network allocation vector (NAV) maintenance, etc.
- Partial BSS Color subfield indicates that the AP considers the BSS color when assigning the association identifiers (AIDs) to STAs joining the BSS. This allows the inclusion of a portion of the BSS color (4 bits out of the 6 bits that is the BSS color) in the PHY header of VHT PPDUs, hence expanding to a certain degree the benefits of the BSS color functionality to these VHT PPDUs.
- BSS Color Disabled subfield indicates whether the use of BSS color is enabled or disabled in the BSS. Since the BSS color is not unique there are cases where multiple BSSs in the neighborhood end up having the same value of the BSS color, namely a BSS color collision, which can lead to undesired MAC behaviors. Hence, the AP can disable the BSS color and related functionalities until the BSS color collision is resolved.

The Basic HE-MCS And NSS Set field indicates the HE-MCSs values for each NSS that are to be supported by all HE STAs in the BSS.

The VHT Operation Information field contains the same subfields as the VHT Operation Information field of the VHT Operation element. The VHT Operation Information field is included when the AP decides to not include the VHT Operation element in the same frame, which can be the case when the AP does not intend to have VHT-only STAs operating in the same BSS as HE STAs in the 5 GHz band.

The Max Co-Hosted BSSID Indicator field contains the maximum number of BSSIDs in a co-hosted BSSID set.

The 6 GHz Operation Information field contains several subfields related to the HE functionalities that are currently in use in the BSS in the 6 GHz band:

- Primary Channel subfield contains the channel number of the primary channel of the BSS.
- Control subfield that contains the Channel Width subfield, which indicates the BSS bandwidth, the Duplicate Beacon subfield, which indicates that beacons

frames are sent in non-HT duplicate mode, and the Regulatory Info subfield, which carries country-specific information related to regulatory rules.

- Channel Center Frequency Segment 0 subfield which indicates a channel center frequency for 20, 40, 80, 160 or 80 + 80 MHz HE BSS.
- Channel Center Frequency Segment 1 subfield which indicates a channel center frequency for 160, or 80 + 80 MHz HE BSS.
- Minimum Rate subfield indicates the minimum rate, in Kbps, that an STA needs to use within this BSS when sending PPDUs (with certain exceptions, e.g. when the PPDU contains a control response frame addressed to the AP).

It is worth noting that the information contained in the Primary Channel, Channel Width, Channel Center Frequency Segment 0, and Channel Center Frequency 1 subfields is the same information that could be obtained from the HT Operation and VHT Operation elements. However, these fields are not allowed in the 6 GHz band since no HT STAs and no VHT STAs are allowed to operate in this band.

7.2.3 Elements Supporting WUR Functionalities

This section describes the IEs transmitted in management frames by the MR to specify the capabilities and operation parameters of the associated WUR STA.

7.2.3.1 WUR Capabilities Element

The WUR Capabilities element advertises support for WUR functionalities. The presence of this element in a transmitted Management frame indicates that all mandatory WUR functionalities are supported by the transmitting WUR device. In addition, the WUR Capabilities element provides a list of optional WUR functionalities supported by the device and other parameters that are related to the device's WUR capabilities.

The format of the WUR Capabilities element is shown in Figure 7.12.

The list of capabilities and capability-related parameters is provided in Table 7.1.

7.2.3.2 WUR Operation Element

The WUR Operation element contains basic WUR parameters that are advertised by a WUR AP and that are essential for setting up and maintaining the WUR functionalities of the associated WUR devices.

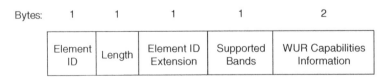

Figure 7.12 WUR Capabilities element.

Table 7.1 Contents of Supported Bands field and WUR Capabilities Information field.

Field name	Definition and description
Supported Bands	Indicates which frequency bands, either 2.4, 5 GHz, or both, are supported by the STA for WUR operation.
	The list of supported frequency bands of the associated STAs is helpful when an AP decides to select a WUR operating channel or when the AP intends to change the operating WUR channel. For example, if all associated STAs that support WUR operation specify that they do support it in the 2.4 GHz frequency band and only a subset of them indicate that they do support it in the 5 GHz frequency band then the AP should select a WUR operating channel that is located in the 2.4 GHz frequency band to ensure that all these STAs can take advantage of this functionality.
	The band that the AP chooses for WUR operation does not necessarily need to be in the same band as the band chosen for MR operation. In this respect, it is worth noting that the 6 GHz frequency band, which is supported by IEEE 802.11ax devices, is not included in the frequency bands that can be supported by the STA for WUR operation (i.e. WUR operation in the 6 GHz frequency band is not defined by the IEEE 802.11ba standard). This way this greenfield band can be exclusively, at least initially, dedicated to MR transmissions, such as transmissions by IEEE 802.11ax devices, that have higher spectrum efficiency compared to WUR transmissions.
Transition Delay	Indicates the maximum amount of time, ranging between 256 and 65 536 μs, that the MR STA requires to transition from doze state to awake state, wherein the transition is caused, for example, by the reception of a WUR Wake-up frame addressed to the WUR STA.
	The transition delay is needed by the STA to essentially turn on all required PHY/MAC components that are expected to be used for transmitting and receiving frames via the MR.
VL WUR Frame Support	Indicates support for variable length (VL) WUR frames.
	By default, a WUR STA is required to support the reception of Fixed Length (FL) WUR frames that have a length of 6 bytes across the board. Similarly, a WUR AP is required to support the transmission of FL WUR frames. These frames are simpler to generate and parse and the duration of the resulting WUR PPDU is 0.924 ms for low data rate (LDR), which is the mandatory WUR data rate.
	The reception and transmission of VL WUR frames are optional due to the extra complexity incurred for creating and parsing variable length frames, since the resulting WUR PPDUs add another eight possible PPDU durations that need support, which vary between 1.18 and 2.972 ms with the mandatory LDR.

(Continued)

Table 7.1 (Continued)

Field name	Definition and description
WUR Group IDs Support	Indicates support for WUR group identifiers and the range of group IDs that are supported by the STA: 1) Between 1 and 16 WUR group IDs, 2) Between 1 and 32 WUR group IDs, 3) Between 1 and 64 WUR group IDs. By default, a WUR STA is required to support only a subset of identifiers that are essential for individual and broadcast WUR operation, such as WUR ID, transmit ID, and non-transmit ID. Functionalities that are related to group-addressed WUR operation are optional, and since storing a list of group IDs requires additional memory, the STA also indicates how many WUR group IDs it can support (see Section 6.4.4).
20 MHz WUR Basic PPDU with HDR Support	Indicates support for WUR Basic PPDU with high data rate (HDR). By default, a WUR STA is required to support the reception of WUR Basic PPDU with LDR, and a WUR AP is required to support the transmission of WUR Basic PPDU with both LDR and HDR. The extra requirement at the AP side derives from the fact that the use of HDR drastically reduces WUR PPDU durations, and hence reducing the channel occupancy. For example, the duration of the PPDU containing an FL WUR frame reduces from 0.924 ms with LDR to 0.284 ms with HDR; and reduces from 2.972 ms with LDR to 0.796 ms with HDR, when the PPDU contains the longest possible VL WUR frame. The reception of WUR Basic PPDUs with HDR, however, is optional for a WUR STA, due to the extra complexity incurred for parsing WUR Basic PPDUs with a different PPDU design (e.g. different SYNC, MC-OOK waveform, etc.) and different PPDU length.
WUR FDMA Support	Indicates support for WUR FDMA operation. By default, a WUR STA is required to support operation in the WUR primary channel where WUR Beacon frames are sent. Operation in WUR secondary channels has additional complexities arising from the fact that the STA needs to dynamically transition between the WUR primary channel and the WUR secondary channel (see Section 6.5).
WUR Short Wake-up Frame Support	Indicates support for WUR Short Wake-up frames. By default, a WUR STA is required to support the use of FL WUR Wake-up frames for the wake-up operation, and optionally support the use of WUR Short Wake-up frames (see Section 6.4.5). WUR Short Wake-up frames provide a more efficient alternative from a channel utilization perspective due to shorter PPDU durations compared to the FL WUR Wake-up frame counterpart (e.g. 0.668 ms as opposed to 0.924 ms for LDR), which is beneficial to reduce the impact in

Table 7.1   (Continued)

Field name	Definition and description
	channel utilization (e.g. channel is already heavily utilized or when WUR Wake-up frames need to be sent with higher frequencies). However, the use of these frames is optional because of the additional complexities that incur from the need of parsing a WUR PPDU containing a shorter Wake-up frame and because of some functionality losses due to the reduced information carried in WUR Short Wake-up frames.

The format of the WUR Operation element is shown in Figure 7.13.

The contents of the WUR Operation Parameters field are shown in Table 7.2.

7.2.3.3   WUR Mode Element

The WUR Mode element is used to set up WUR operation and exchange the necessary parameters that are needed for enabling and assisting various WUR functionalities.

The format of the WUR Mode element is shown in Figure 7.14.

The Action Type field provides the type of action for the WUR mode and eventually whether this is a request or a response during the negotiation. The action types fall into three categories:

- Request: Used by a WUR non-AP STA during negotiation for requesting to operate in WUR Mode or for requesting to suspend an existing WUR Mode.
- Response: Used by the WUR AP during negotiation for responding to received requests for either setting up WUR Mode or suspending an existing WUR Mode.
- Transition: Used by a WUR non-AP STA during WUR operation to suspend an existing WUR mode or to resume a previously suspended WUR Mode.

The WUR Mode Response Status field indicates the outcome of the negotiation, and in the case of an unsuccessful negotiation, it provides the reason for the failed negotiation (if any).

Table 7.3 lists the possible outcomes (if any) provided by the WUR Mode Response Status field depending on the action type and a description of their usage.

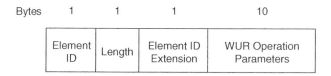

Figure 7.13   WUR Operation element format.

Table 7.2 Contents of the WUR Operation Parameters field.

Field name	Definition
Minimum Wake-up Duration	Indicates the minimum duration that the WUR AP intends to use for any WUR service periods (SPs) that are to be negotiated between a WUR STA and the WUR AP (see Section 6.4.3). This field has a value that varies between 0 and 65 280 µs, with a granularity of 256 µs.
Duty Cycle Period Unit	Indicates the basic unit that the WUR AP intends to use for any WUR duty cycle periods that are to be negotiated by WUR STAs with the WUR AP (see Section 6.4.3). This field has a value that varies between 0 and 262 140 µs, with a granularity of 4 µs.
WUR Operating Class	Indicates the operating class for the WUR primary channel that the AP uses for the WUR BSS (see Section 6.3.2). The operating class is interpreted in the context of the country, which is indicated in the Country element contained in the same Beacon frame that contains the WUR Operation element.
WUR Channel	Indicates the channel number of the WUR primary channel of the WUR BSS, which is interpreted in the context of the WUR operating class.
WUR Beacon Period	Represents the periodicity of WUR Beacon frame transmissions, i.e. the interval of time that separates the target transmission times of two consecutive WUR Beacon frames (see Section 6.3.2). This field has a value that varies between 0 and 65 535 time units (TUs), where the TU is equal to 1024 µs. The value 0 has no practical use since it implies that the AP is scheduling for transmitting all WUR Beacons at the same time. Similarly, other low values, e.g. 1 TU, 2 TU, etc., have limited practical use since the AP would be using a significant amount of airtime just for transmitting WUR Beacon frames (e.g. ~100% for 1 TU, ~50% for 2 TU and so on).
Offset of TWBTT	The Offset Of TWBTT subfield indicates the time difference, in units of TU, between the first occurring target WUR beacon transmission time (TWBTT) and the start of the timing synchronization function (TSF) timer (i.e. the TSF timer value of 0) (see Section 6.3.2).
Counter	Indicates the most recent value of the Counter subfield that is included in broadcast WUR Wake-up frames that are sent under the critical BSS update delivery context (see Section 6.4.4). This field has an integer value that varies between 0 and 15.
Common PN	A bit that is set to 1 to indicate that all protected WUR frames generated within the WUR BSS share a common packet number (PN) and is set to 0 to indicate that protected WUR frames addressed to different receivers use independent PNs (see Section 6.6).

Bytes:	1	1	1	1	1	1	Variable
	Element ID	Length	Element ID Extension	Action Type	WUR Mode Response Status	WUR Parameters Control	WUR Parameters

Figure 7.14 WUR Mode element.

Table 7.3 Combined settings for Action Type and WUR Mode Response Status fields.

Action Type	WUR Mode Response Status	Description
Enter WUR Mode Request	Invalid	The WUR non-AP STA is requesting to enter the WUR mode.
		The WUR Mode Response Status field, in this case, is invalid, and as such its value is ignored by the receiving WUR AP.
Enter WUR Mode Response	Accepted	The WUR AP responds that it has accepted the request of the WUR non-AP STA to enter the WUR mode.
	Denied due to unspecified reason	The WUR AP responds that the request of the WUR non-AP STA to enter the WUR mode is rejected for no specific reason.
		Given that the WUR AP has not given a specific reason for which it has rejected the request from the STA to enter WUR mode, it is likely that the AP will continue to reject subsequent requests that the STA might send to the AP, despite any changes that the STA makes to the parameters that it intends to use during the requested WUR mode.
	Denied because the preferred duty cycle period is too large.	The WUR AP responds that the request of the WUR non-AP STA to enter the WUR mode is rejected because the specified duty cycle period is too large.
		In this case the WUR AP has given a reason for the rejection, however, the information provided to the STA is somewhat limited since the STA does not know what value of the preferred duty cycle period would be acceptable for the WUR AP.
		The WUR STA may continue to send subsequent requests to the AP while gradually reducing the preferred duty cycle period that it indicates in each of these requests until one of the requests is accepted by the WUR AP.

(Continued)

Table 7.3 (Continued)

Action Type	WUR Mode Response Status	Description
	Denied because the WUR AP cannot satisfy the request of generating keep-alive WUR frames every WUR duty cycle service period.	The WUR AP responds that the request of the WUR non-AP STA to enter the WUR mode is rejected because the WUR AP cannot generate keep-alive WUR frames.
		In this case, the WUR AP has given a very specific reason for the rejection.
		The WUR STA may send a subsequent request to the AP that is very likely to be accepted by the WUR AP if the request does not include the generation of keep-alive WUR frames.
Enter WUR Mode Suspend Request	Invalid	The WUR non-AP STA is requesting to enter WUR mode suspend.
		The WUR Mode Response Status field, in this case, is invalid, and as such its value is ignored by the receiving WUR AP.
Enter WUR Mode Suspend Response	Accepted	The WUR AP responds that it has accepted the request of the WUR non-AP STA to enter WUR mode suspend.
	Denied due to unspecified reason.	The WUR AP responds that the request of the WUR non-AP STA to enter WUR mode suspend is rejected for no specific reason.
		Given that the WUR AP has not given a specific reason for which it has rejected the request from the STA to enter WUR mode suspend, it is likely that the AP will continue to reject subsequent requests that the STA might send to the AP.
	Denied because the preferred duty cycle period is too large.	The WUR STA may continue to send subsequent requests to the AP while gradually reducing the preferred duty cycle period that it indicates in each of these requests until one of the requests is accepted by the WUR AP.

Table 7.3   (Continued)

Action Type	WUR Mode Response Status	Description
	Denied because the WUR AP cannot satisfy the request of generating keep-alive WUR frames every WUR duty cycle service period.	The WUR AP responds that the request of the WUR non-AP STA to enter the WUR mode is rejected because the WUR AP cannot generate keep-alive WUR frames.
		In this case, the WUR AP has given a very specific reason for the rejection.
		The WUR STA may send a subsequent request to the AP that is very likely to be accepted by the WUR AP if the request does not include the generation of keep-alive WUR frames.
Enter WUR Mode Suspend	Invalid	The WUR non-AP STA is suspending the current WUR mode.
Enter WUR Mode	Invalid	The WUR non-AP STA is resuming the WUR mode.

The contents of the WUR Parameter Control field and the WUR Parameters field depend on the settings of the preceding Action Type field, WUR Mode Response Status field and specifically the device that generates the WUR Mode element. Both these fields provide valid information only if:

- When the WUR Mode element is sent by a WUR AP then the Action Type field is either Enter Mode Response or Enter WUR Mode Suspend Response and the WUR Mode Response Status is Accepted. Both fields are reserved (i.e. ignored upon reception) for any other combination.
- When the WUR Mode element is sent by a WUR non-AP STA then both fields provide valid information for any valid value of Action Type field (noting that in the case of a WUR non-AP STA the contents of the WUR Mode Response Status are not valid).

The WUR Parameter Control contains fields that indicate the presence or absence of certain fields in the subsequent WUR Parameters field:

- WUR Duty Cycle Start Time Present, which indicates whether the WUR Duty Cycle Start Time subfield in the WUR Parameters field is present or not. This field is set to 0 by a WUR non-AP STA since the WUR Duty Cycle Start Time is not present in a WUR Mode element sent by a WUR non-AP STA (see Table 7.4).

Table 7.4 Contents of the WUR Parameters field for a WUR non-AP STA.

Subfield Name	Definition and description
WUR Duty Cycle Service Period	Indicates the preferred WUR duty cycle service period during which the WUR STA expects to be in WUR awake state (see Section 6.4.3). This field has a value that varies between 0 and 17 179 869 180 µs (~4.8 h), with a granularity of 4 µs.
Duty Cycle Period	Indicates the preferred elapsed time between the start times of two successive WUR duty cycle service periods (see Section 6.4.3). This field has a value that varies between 0 and 17 179 344 900 µs (~4.8 h), with a granularity that is provided in the Duty Cycle Period Units field of the WUR Operation element.
	Note that the granularity is provided by the WUR AP in the WUR Operation element, and hence the WUR STA can only choose a duty cycle period that is within the possible range that the WUR AP allows. For example, if the WUR AP chooses a duty cycle period unit of 256 µs, then the WUR STA can only chooses a duty cycle period between 0 and 16 776 960 µs (~16.8 s).
Proposed WUR Parameters	The Proposed WUR Parameters field is only present if the Proposed WUR Parameters Present subfield in the WUR Parameters Control field is 1. The absence of the Proposed WUR Parameters field from a transmitted WUR Mode element is an explicit indication to the AP that the WUR STA does not have any recommendation or request for the parameters listed below.
	Recommended WUR (Short) Wake-up Frame Rate Indicates whether a certain data rate (LDR or HDR) is recommended by the STA to the AP for use when sending WUR (Short) Wake-up frames to this STA. The value 0 indicates that there is no recommendation by the STA, the value 1 indicates that LDR is recommended and the value 2 indicates that HDR is recommended.
	Recommended WUR Channel Offset Indicates the recommended location of the WUR channel relative to the WUR primary channel.
	The Recommended WUR Channel Offset field specifies a value of 0 to indicate that the recommended WUR channel has the same frequency location as the WUR primary channel while the value 7 is used to indicate that the STA does not have any recommendation for the location of the WUR channel.

Table 7.4 (Continued)

Subfield Name	Definition and description
	Alternatively, the Recommended WUR Channel Offset field contains an odd natural value to indicate that the recommended WUR channel is higher in frequency compared to the WUR primary channel and contains an even natural value to indicate that the recommended WUR channel is lower in frequency compared to the WUR primary channel. More precisely, the recommended WUR Channel Offset field specifies: • A value of 1, 3, and 5 to indicate that the recommended WUR channel is the first, second, and third 20 MHz channel, respectively, above the WUR primary channel, • A value of 2, 4, and 6 to indicate that the recommended WUR channel is the first, second, and third 20 MHz channel, respectively, below the WUR primary channel.
Requested Keep-Alive WUR Frame	Indicates that the WUR STA is requesting the AP to transmit keep-alive WUR frames during each WUR duty cycle service period that the STA has negotiated with the AP (see Section 6.4.6).

- WUR Group ID List Present, which indicates whether the WUR Group ID List subfield in the WUR Parameters field is present or not. Similarly, this field is set to 0 by a WUR non-AP STA since the WUR Group ID List field is not present in a WUR Mode element sent by a WUR non-AP STA (see Table 7.4).
- Proposed WUR Parameters Present, which indicates whether the Proposed WUR Parameters subfield in the WUR Parameters field is present or not. In this case, this field is set to 0 by a WUR AP since the Proposed WUR Parameters field is not present in a WUR Mode element sent by a WUR AP (see Table 7.5).

The contents of the WUR Parameters field for a WUR non-AP STA are shown in Table 7.4.

The contents of the WUR Parameters field for a WUR AP STA are shown in Table 7.5.

Table 7.5 Contents of the WUR Parameters field for a WUR AP.

Subfield name	Definition and description
WUR ID	Indicates the WUR identifier (ID), which uniquely identifies the WUR non-AP STA within the WUR BSS.
	The WUR ID is assigned to the WUR non-AP STA by the WUR AP and may appear in certain types of WUR frames addressed to the WUR non-AP STA (see Section 7.4).
WUR Channel Offset	Indicates the location of the WUR channel relative to the WUR primary channel. The WUR channel is expected to be used by the WUR AP to transmit certain WUR frames addressed to the WUR non-AP STA (see Section 6.5).
	The WUR Channel Offset field specifies a value of 0 to indicate that the WUR channel has the same frequency location as the WUR primary channel.
	The WUR Channel Offset field specifies an odd natural value to indicate that the WUR channel is higher in frequency compared to the WUR primary channel and specifies an even natural value to indicate that the WUR channel is lower in frequency compared to the WUR primary channel. More precisely, the WUR Channel Offset field specifies:
	• A value of 1, 3, and 5 to indicate that the WUR channel is the first, second, and third 20 MHz channel, respectively, above the WUR primary channel,
	• A value of 2, 4, and 6 to indicate that the WUR channel is the first, second, and third 20 MHz channel, respectively, below the WUR primary channel.
	These WUR Channel Offset values cover all possible WUR channel locations within an 80 MHz bandwidth, relative to a WUR primary channel, which on its own can be located anywhere within that same 80 MHz bandwidth.
WUR Duty Cycle Start Time	The WUR Duty Cycle Start Time is only present if the WUR Duty Cycle Start Time Present subfield in the WUR Parameters Control field is 1. The WUR Duty Cycle Start Time is not present only when the WUR STA is requesting to operate in WUR mode without any duty cycle (i.e. the WUR STA intends to be in WUR awake state continuously).
	If the WUR Duty Cycle Start Time is present, then it provides the time at which the WUR AP intends to initiate WUR Duty Cycle operation for the receiving WUR non-AP STA (see Section 6.4.3).
	The time is provided in the form of a positive integer that corresponds to the value that the TSF time will have at that specific point in time and has a resolution of one microsecond.

Table 7.5 (Continued)

Subfield name	Definition and description
WUR Group ID List	The WUR Group ID List is only present if the WUR Group ID List Present subfield in the WUR Parameters Control field is 1. The absence of the WUR Group ID List from a transmitted WUR Mode element is an explicit indication that the WUR AP has not allocated any WUR Group IDs to the receiving WUR non-AP STA.
	If the WUR Group ID List is present, then it provides a list of WUR group IDs that are assigned by the WUR AP to the receiving WUR non-AP STA. Any of these WUR Group IDs may appear in certain types of WUR frames addressed to the WUR non-AP STA (see Section 6.4.4).
	The WUR Group ID List is provided in the form of a Starting WUR Group ID and a WUR Group ID Bitmap as defined below.

	WUR Group ID Bitmap Size	Specifies the size of the WUR Group ID Bitmap field in the WUR Group ID List.
		The WUR Group ID Bitmap Size is set to 0 to indicate that the size of the WUR Group ID Bitmap subfield is zero, i.e. the WUR Group ID Bitmap subfield is not present.
		The WUR Group ID Bitmap Size is set to 1, 2, and 3 to indicate that the size of the WUR Group ID Bitmap field is 16 bits, 32 bits, and 64 bits, respectively.
	Starting WUR Group ID	The definition of this field depends on the value of the WUR Group ID Bitmap Size.
		If the WUR Group ID Bitmap Size is 0 then the Starting WUR Group ID contains a single WUR group ID assigned by the WUR AP to the WUR non-AP STA.
		If the WUR Group ID Bitmap size is greater than 0 then the Starting WUR Group ID contains the starting group ID of the WUR Group ID Bitmap subfield.
	WUR Group ID Bitmap	Specifies the WUR Group IDs assigned by the WUR AP to the WUR non-AP STA. The WUR group ID bitmap starts with the group ID identified by the value of the Starting WUR Group ID subfield, $\text{Gid}_{\text{start}}$, and ends with the group ID identified by $\text{Gid}_{\text{end}} = (\text{Gid}_{\text{start}} + \text{Bitmap size}) \bmod 4096$
		Within the WUR group ID bitmap, bit position n is set to 1 to indicate that the WUR group ID, $\text{Gid}_n = (\text{Gid}_{\text{start}} + n) \bmod 4096$ is assigned to the STA and is set to 0 to indicate that Gid_n is not assigned to the STA.

7.2.3.4 WUR Discovery Element

The WUR Discovery element advertises a set of WUR discovery channels and may contain additional information that can help a WUR STA in discovering WUR APs that operate in the area.

The format of the WUR Discovery element is shown in Figure 7.15.

The WUR Discovery element contains a number, n, of WUR AP Information fields. Each WUR AP Information field contains:

- The WUR Discovery Operating Class and the WUR Discovery Channel, which indicate the operating class and the channel in use by all WUR APs that are being reported by this WUR AP Information field,
- The total number, m, of WUR APs reported by this WUR AP Information field is provided in the WUR AP Count subfield and for each reported AP there is a WUR Parameter field that includes a list of WUR AP parameters.

Each WUR AP Parameter field contains:

- A WUR AP Parameters Control subfield that indicates whether the transmitting WUR AP is advertising its own WUR discovery information or the WUR discovery information of other WUR APs. It additionally contains indications of the presence of the Short SSID, BSSID, and WUR Discovery Period subfields within the corresponding WUR AP Parameter field.
- A Short SSID subfield that indicates the short SSID of the network of the reported WUR AP,
- A BSSID subfield that indicates the BSSID of the reported WUR AP,
- A WUR Discovery Period subfield that indicates the periodicity of transmitted WUR Discovery frames by the reported WUR AP.

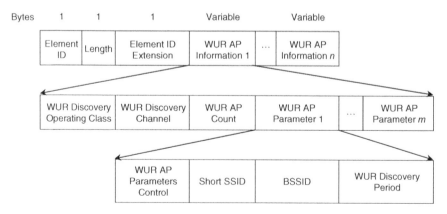

Figure 7.15 WUR Discovery element format.

Bytes: 1   1   1   1   0 or 6

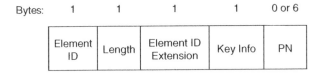

Element ID	Length	Element ID Extension	Key Info	PN

Figure 7.16 WUR PN Update element format.

7.2.3.5 WUR PN Update Element

The WUR PN Update element is used by the WUR AP to update the packet number (PN) that is maintained by a WUR non-AP STA. The PN may need to be updated for several reasons, which are described in Section 6.6.

The format of the WUR PN Update element is shown in Figure 7.16.

The Key Info field contains the Key ID corresponding to the WUR temporal key (WTK) or WUR integrity temporal key (WIGTK), and a PN Present subfield that indicates whether the PN field is present or not.

The PN field, if present, contains the PN that is to be used by the receiver to update the local PN for WUR Wake-up frames.

7.3 Main Radio MAC Frames

IEEE 802.11 has defined over the years many MAC frames. The general frame format for MAC frames is shown in Figure 7.17.

IEEE 802.11ba uses several of these MAC frames, in particular management frames, to carry the IEs defined in Section 7.2. In the Section 7.3.1, we provide a brief summary of the main management frames that are exchanged between a WUR AP and WUR STAs.

7.3.1 Beacon Frame

The Beacon frame is a management frame that contains all the required information about the network. These frames are sent by the AP periodically and, in addition to announcing the presence of the network, also aid associated STAs in synchronizing their clocks.

Bytes: 2   2   6   0 or 6   0 or 6   0 or 2   0 or 6   0 or 2   0 or 4 Variable 4

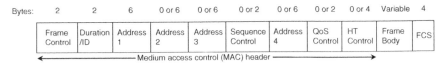

Frame Control	Duration /ID	Address 1	Address 2	Address 3	Sequence Control	Address 4	QoS Control	HT Control	Frame Body	FCS

◄─────────────────────── Medium access control (MAC) header ───────────────────────►

Figure 7.17 MAC frame format.

The Beacon frame contains an ever-increasing list of elements that provide information for a variety of functionalities supported by the network and other operation parameters that are needed for operating as a member of the network. A WUR AP adds the following elements to the Beacon frame body:

- WUR Capabilities element to advertise that the AP supports WUR operation
- WUR Operation element to advertise operational parameters currently in use by the WUR AP
- WUR Discovery element to advertise information beneficial for STAs that are scanning for WUR APs in the area.

The Beacon frame is additionally used to advertise the presence of downlink data for specific WUR STAs, for which the STAs need to wake the MR in order to be able to poll the AP and retrieve the data (see Section 6.4).

7.3.2   Probe Request/Response Frames

Probe Request/Response frames are management frames that are used for active scanning and as such contain useful information to discover the presence of networks in the area and information relevant to operating in those networks.

Probe Request frames are sent by a WUR non-AP STA and from a WUR perspective they only contain the WUR Capabilities element, which notifies the APs that are responding to the Probe Request frames that the specific STA supports WUR functionalities. Probe Response frames are sent by a WUR AP and from a WUR perspective they contain the same IEs as the Beacon frame (i.e. WUR Capabilities, WUR Operation, and WUR Discovery elements).

7.3.3   (Re)Association Request/Response Frames

An STA needs to associate with an AP before using the services provided by that AP. Association is achieved by sending an Association Request frame to the AP and receiving in return an Association Response frame. The STA can also re-associate with an AP while already associated. Re-association process uses equivalent management frames that are called Reassociation Request and Reassociation Response frames.

The (Re)Association Request frame contains the WUR Capabilities element and can additionally contain the WUR Mode element. The WUR STA includes the WUR Mode element if the WUR STA wants to both associate and setup WUR mode with the WUR AP. An Association Request frame that contains a WUR Mode element is a WUR Mode Request frame for the purposes of WUR operation.

The (Re)Association Response frame contains the WUR Capabilities element, the WUR Operation element and can additionally contain the WUR Mode element.

The WUR AP includes the WUR Mode element in the (Re)Association Response frame if the soliciting (Re)Association Request frame sent by the WUR STA contained a WUR Mode element as well. That is, the WUR STA wants to both associate and setup WUR mode with the WUR AP during the (re)association phase. An Association Response frame that contains a WUR Mode element is a WUR Mode Response frame for the purposes of WUR operation.

7.3.4 Action Frames

Action frames are management frames that are designed to provide a specific functionality (or action).

IEEE 802.11ba has defined three different types of action frames:

1) WUR Mode Setup frame – Can be used for WUR mode negotiation and for updating WUR mode parameters. When sent by a WUR non-AP STA the WUR Mode Setup frame contains the WUR Mode element, and acts as a WUR mode request. When sent by a WUR AP the WUR Mode Setup frame contains the WUR Mode and the WUR Operation elements and may additionally contain the WUR PN Update element. In this case, it acts as a WUR mode response.
2) WUR Mode Teardown frame – Can be used to tear down an existing WUR mode
3) WUR Wake-up Indication frame – Can be used to cause an unsolicited wake-up.

7.4 WUR MAC Frames

IEEE 802.11ba has defined several WUR frames with different functionalities and purposes.

All WUR frames share some common components (e.g. Frame Control, ID, and FCS fields). Other components (such as Type Dependent (TD) Control, Frame Body fields) are present only in certain WUR frame types and carry information that is specific to the type of WUR frame.

The frame format for WUR frames is shown in Figure 7.18.

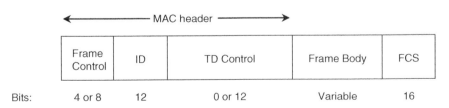

Figure 7.18 Wake-up Radio (WUR) frame format.

The Frame Control field identifies the WUR frame type and additionally contains information relevant to the WUR frame construction, enabling the receiver to process and correctly decode the WUR frame. The Frame Control field contains:

- A 3-bit Type subfield that indicates the type of the WUR frame. IEEE 802.11ba has defined five different types of WUR frames. The remaining three values of the Type subfield are reserved and left for future use.
- A 1-bit Protected subfield which indicates whether the WUR frame is protected with a message integrity check (MIC) algorithm (see Section 6.6) or with a cyclic redundancy check (CRC).
- A 1-bit Frame Body Present subfield that indicates whether the Frame Body field is present in the WUR frame or not. The WUR frame is an FL WUR frame when this bit is 0 and is a Variable Length (VL) WUR frame when this bit is 1.
- A 3-bit Length/Miscellaneous subfield that indicates the length of the Frame Body field if the Frame Body is present, where the length varies between 2 bytes and 16 bytes. If the Frame Body field is not present, then this field contains miscellaneous information and depends on the type of the WUR frame.

Table 7.6 summarizes the WUR frame types and lists the possible frame lengths with respective WUR PPDU durations.

The ID field contains an identifier for the WUR frame. These identifiers are defined in Table 7.7.

Table 7.6 Summary of WUR frame types.

Type	Can the frame be protected?	Can the frame body be present?	WUR frame length (bytes)	WUR PPDU duration (ms)
WUR Beacon	No	No	6	LDR: 0.924 HDR: 0.284
WUR Wake-up	Yes	Yes	6–24	LDR: 0.924–2.972 HDR: 0.284–0.796
WUR Vendor Specific	Yes	Yes	6–24	LDR: 0.924–2.972 HDR: 0.284–0.796
WUR Discovery	No	Yes	10	LDR: 1.436 HDR: 0.412
WUR Short Wake-up	No	No*	4	LDR: 0.668 HDR: 0.220

Table 7.7 Identifiers in ID field of WUR frames.

Identifier	Definition and description
Transmitter ID	Identifier of the transmitting AP.
	The transmitter ID is obtained from the 12 least significant bits (LSBs) of the compressed BSSID.
	The transmitter ID is used in broadcast WUR frames that are sent by WUR AP on behalf of the transmitted BSSID to all WUR non-AP STAs that are associated to the transmitted BSSID.
Non-transmitter ID	Identifier of the non-transmitted BSSID.
	The non-transmitter ID is obtained from the transmitted BSSID + k, where k is the BSSID index of the non-transmitted BSSID.
	The non-transmitter ID is used in broadcast WUR frames that are sent by the WUR AP on behalf of a non-transmitted BSSID to all WUR non-AP STAs that are associated to the non-transmitted BSSID.
WUR Group ID	Identifier of a group of receiving WUR non-AP STAs.
	The WUR group ID is used in group cast WUR frames that are sent by the WUR AP to a group of WUR non-AP STAs.
WUR ID	Identifier of an individual receiving WUR non-AP STA.
	The WUR ID is used in unicast WUR frames that are sent to by the WUR AP to a specific WUR non-AP STA that is assigned this WUR ID.
OUI1	Identifier that is obtained from the Organizationally Unique Identifier (OUI).

Both the TD Control and the Frame Body fields, if present, contain information that is specific to the type of the WUR frame and are covered in more detail in the rest of this section.

The FCS field contains a 16-bit cyclic redundancy check (CRC) or a 16-bit message integrity check (MIC) depending on the value of the Protected subfield of the Frame Control field.

The CRC helps the receiver in determining whether the received WUR frame is received correctly or not while the MIC additionally helps the receiver determining whether the received WUR frame is actually sent by the intended transmitter rather than an attacker.

The CRC is calculated over all the fields that are present in the WUR frame and additionally over an Embedded BSSID field, with the Embedded BSSID field being present only for certain types of WUR frames, which are described in more detail in the remaining subsections. The schematic for the computation of the CRC is shown in Figure 7.19.

While the Embedded BSSID is part of the calculation fields, it is actually not included in the WUR frame that is transmitted over the air. Hence, a receiving WUR STA can decode the WUR frame correctly only if it knows the Embedded BSSID that is being used by the WUR AP that is sending the WUR frame. The WUR STA can obtain the Embedded BSSID that is being used by the WUR AP by extracting the 16 MSBs of a 32-bit CRC calculated based on the BSSID contained in Beacon frames transmitted by the WUR AP or the transmitted BSSID of the multiple BSSID set.

The CRC is computed as the 1s complement of the remainder generated by the modulo 2 division of the calculation fields, by the polynomial $x^{16}+x^{12}+x^5+1$, where the shift register state is preset to all 1s and is shown in Figure 7.20, where the Serial Data Input is the calculation fields and the Serial Data Output is the computed 16-bit CRC. The transmitter then includes the obtained 16-bit CRC to the FCS field of the WUR frame that is being sent, while excluding the Embedded BSSID field. At the other end, the receiver performs the same CRC computation

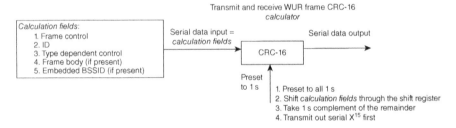

Figure 7.19 Schematic for 16-bit CRC computation.

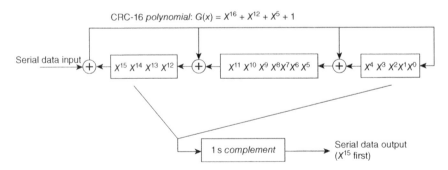

Figure 7.20 CRC-16 implementation.

over the received *calculation fields* and if the obtained 16-bit CRC is the same as the 16-bit CRC that is included in the FCS of the received WUR frame then the WUR frame is received correctly. If there is at least one bit mismatch, then the frame is not received correctly.

The rest of this section describes more details about the five types of WUR frames, which are derived from the general WUR frame format shown in Figure 7.18.

7.4.1 WUR Beacon Frame

WUR Beacon frames contain specific information for a WUR BSS. A WUR AP sends these frames to maintain connectivity with associated WUR STAs and help the WUR STAs in synchronizing their TSF timer.

The format of the WUR Beacon frame is shown in Figure 7.21.

The Frame Control field indicates that the frame is not protected and that the Frame Body field is not present. This ensures that any associated WUR STA can correctly decode a WUR Beacon frame without the need of supporting additional capabilities, such as support for the reception of protected WUR frames and/or support for the reception of VL WUR frames.

The WUR Beacon frame is a broadcast frame and as such the ID field is set to the transmit ID.

The TD Control field of the WUR Beacon frame contains a 12-bit partial time stamp function (TSF) timer that helps associated WUR STAs to synchronize their TSF timers to that of the WUR AP when they receive the WUR Beacon frame. The TSF synchronization mechanism is explained in more detail in Section 6.3.

The FCS field contains the CRC, wherein the *calculation fields* also includes the Embedded BSSID field (i.e. these frames can be received successfully only by a WUR STA that already knows the BSSID that is in use by the WUR BSS).

7.4.2 WUR Wake-up Frame

WUR Wake-up frames are used to wake up the MR of one or more of the WUR STAs. These frames contain additional information useful for the receiving WUR

Frame Control	ID (Transmit ID)	TD Control (Partial TSF)	FCS (CRC)
8	12	12	16

Figure 7.21 WUR Beacon frame format.

STAs, such as indications of the presence at the MR of critical BSS updates and/or the presence of group addressed BUs, partial TSF timer information for clock synchronization, etc.

The WUR Wake-up frame comes in both variants, FL WUR Wake-up frame, and VL WUR Wake-up frame.

The format of the FL WUR Wake-up frame is shown in Figure 7.22, and the format of the VL WUR Wake-up frame is shown in Figure 7.23.

The Frame Control field indicates whether the frame is protected or not, and whether the frame contains the Frame Body field or not (i.e. essentially differentiating between a VL WUR Wake-up and an FL WUR Wake-up frame). In addition, the Frame Control field of an FL WUR Wake-up frame contains the Miscellaneous subfield, which contains two bits:

- Group Addressed BU: Set to 1 to indicate that group addressed frames are buffered at the AP corresponding to the BSSID identified by the ID field and is set to 0 to indicate that no group addressed frames are buffered at the AP. This bit's setting is valid only in broadcast FL WUR Wake-up frames because such indication is of interest for all WUR STAs that are members of that BSSID (see Section 6.4.4). The bit is reserved (i.e. set to 0 and ignored upon reception) in all other FL WUR Wake-up frames (individually addressed or group addressed).
- Key ID: Set to 1 to indicate that the WIGTK identifier of the WIGTK used to compute the MIC is 8 and is set to 0 to indicate that the WIGTK identifier of the WIGTK used to compute the MIC is 9. This bit's setting is valid only in FL WUR Wake-up frames that are protected by WIGTK (i.e. broadcast and group addressed FL WUR Wake-up frames as described in Section 6.6). The bit is reserved in all other FL WUR frames (unprotected or individually addressed).

Frame Control	ID	TD Control	FCS (CRC or MIC)
8	12	12	16

Figure 7.22 FL WUR Wake-up frame format.

Frame Control	ID (WUR Group ID)	TD Control	Frame Body (STA Info List)	FCS (CRC or MIC)
8	12	12	Variable	16

Figure 7.23 VL WUR Wake-up frame format.

The ID field setting depends on the variant of the WUR Wake-up frame:

- For FL WUR Wake-up frame: Set to a WUR ID if the frame is individually addressed, to a WUR group ID if the frame is group addressed and set to transmit ID or non-transmit ID if the frame is broadcast.
- For VL WUR Wake-up frame: Set to a WUR group ID that is known by all STAs that support the reception of the VL WUR Wake-up frame, and the list of WUR IDs for the STAs that belong to this specific WUR group ID is included in the Frame Body field of the frame.

The flexibility of having all these different ID settings for WUR Wake-up frames enables the AP to wake up a single STA by using the WUR ID, a group of STAs by using a WUR group ID, and a subset of a group of STAs by including the list of WUR IDs in the case of a VL WUR Wake-up frame, and all associated STAs by using the transmit ID or the non-transmit ID.

The TD Control field contains the following two subfields:

- Sequence Number subfield, which contains an 8-bit partial PN, namely PN0, when the WUR Wake-up frame is protected (see Section 6.6) and is reserved otherwise.
- Counter subfield which contains a 4-bit BSS update counter if the frame is a broadcast WUR Wake-up frame, contains 4-bits of PN1 if the frame is not broadcast and some other protection-related conditions are satisfied (see Section 6.6), and is reserved otherwise. The BSS update counter is an unsigned integer that is initialized to 0 and then increments every time that a critical update has occurred to the BSS parameters (see Section 6.4.4).

The Frame Body field is only present in the WUR VL Wake-up frame (i.e. when the Frame Body Present field is set to 1) and contains a list of n STA Info fields, where n can be at most 8 because of the 16 bytes length limitation of the Frame Body field.

Each STA Info field is addressed to a specific WUR STA (identified by the WUR ID subfield in the respective STA Info field), and contains a reserved subfield, which is left for future use. The format of the Frame Body field is shown in Figure 7.24.

Figure 7.24   Format of the Frame Body field in a WUR VL Wake-up frame.

The FCS field contains the CRC if the frame is unprotected and the MIC if the frame is protected. When the FCS field contains the CRC, the *calculation fields* include the Embedded BSSID field as well (i.e. these frames can be received successfully by a WUR STA that already knows the BSSID that is in use by the WUR BSS).

7.4.3 WUR Discovery Frame

WUR Discovery frames contain the minimum amount of network information that is needed for discovering a specific network. A WUR AP sends these frames to advertise the network's presence to WUR STAs in the area and provide preliminary information related to the advertised network, which can be used by a receiving WUR STA to determine the BSSID, the SSID, and the operating channel of the advertised network, as described in Section 6.2.

The format of the WUR Discovery frame is shown in Figure 7.25.

The Frame Control field indicates that the frame is not protected, since these frames are expected to be received by any WUR STA that is within the coverage area of the WUR AP, and that are not necessarily associated to the AP, and without the need of any password or security key.

The ID field and the TD Control field contain the compressed BSSID, which is equally split between the ID field, which carries the 12 least significant bits (LSBs) of the compressed BSSID (i.e. the transmitter ID), and the TD Control field, which carries the 12 most significant bits (MSBs).

The Compressed SSID field contains the 16 LSBs of the short SSID of the network that is being advertised.

The Operating Class And Channel field indicates the operating class and the location of the primary channel of the BSS that is being advertised.

The FCS field contains the CRC, wherein the *calculation fields* do not include the Embedded BSSID field (i.e. these frames can be received successfully by any WUR STA without knowing the BSSID that is in use by the WUR BSS).

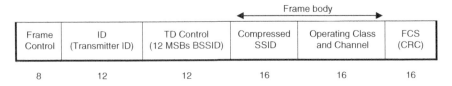

Figure 7.25 WUR Discovery frame format.

7.4.4 WUR Vendor-Specific Frame

WUR Vendor-Specific frame contains information that is not defined by the IEEE 802.11ba standard but is instead defined by each vendor (which can be an organization or company). These frames allow vendors to provide supplementary information that may enhance WUR functionalities and/or address additional applications or use cases that were not in scope of IEEE 802.11ba at the time of standardization.

The format of the WUR Vendor-Specific frame is shown in Figure 7.26.

The Frame Control field indicates whether the frame is protected or not, and additionally whether the Frame Body is present in which case it provides its length as well.

The ID field and the TD Control field contain the Organizationally Unique Identifier (OUI), which is 24-bit number that is uniquely assigned to a vendor, manufacturer, or organization.

The 24-bit OUI is equally split between the ID field, which carries the 12 least significant bits (LSBs), and the TD Control field, which carries the 12 most significant bits (MSBs).

A list of sample OUIs (in their hexadecimal representation), that were selected from the OUIs assigned to organizations, whose affiliated members participated one way or the other in the development of IEEE 802.11ba standards, is provided in alphabetical order in Table 7.8. It is worth noting that while these OUIs are uniquely assigned to an organization there are several instances wherein a particular organization may have multiple assigned OUIs to it. This enables an organization that has multiple OUIs assigned to it to create different variants of WUR Vendor-Specific frames, each of which identified by its own OUI, and each of which for a specific purpose or application.

The vendor-specific information in a WUR Vendor-Specific frame is carried either in the Miscellaneous subfield of the Frame Control field in the case of an FL WUR frame or in the Frame Body field in the case of a VL WUR frame.

The FCS field carries either the CRC or the MIC, however, the protection algorithms that are used to protect the frame are vendor specific and outside the scope of the IEEE 802.11ba standard.

Frame Control	ID (12 LSBs OUI)	TD Control (12 MSBs OUI)	Frame Body	FCS (CRC or MIC)
8	12	12	0 to 16	16

Figure 7.26 WUR Vendor-Specific frame format.

Table 7.8 List of sample OUIs.

	OUI	Organization		OUI	Organization
1	B8-C1-11	Apple, Inc.	9	00-0C-E7	MediaTek Inc.
2	1C-69-A5	BlackBerry RTS	10	A4-81-EE	Nokia Corporation
3	34-07-FB	Ericsson AB	11	08-00-23	Panasonic Comm. Co., Ltd.
4	70-10-6F	Hewlett Packard Enterprise	12	64-9C-81	Qualcomm Inc.
5	48-AD-08	Huawei Technologies CO., LTD	13	00-23-C2	Samsung Electronics. Co. LTD
6	00-03-47	Intel Corp.	14	78-D6-B2	Toshiba
7	5C-17-D3	LGE	15	74-A7-8E	ZTE Corporation
8	00-50-43	Marvell Semiconductor, Inc.			

Source: https://standards-oui.ieee.org/.

7.4.5 WUR Short Wake-up Frame

WUR Short Wake-up frames provide a reduced set of functionalities compared to the WUR Wake-up frames described in Section 7.4.2, namely they are only used to wake up the MR of a single WUR STA as described in Section 6.4.5. As the name states these are shorter frames (4 bytes long compared to the minimum of 6 bytes long of FL WUR Wake-up frame), which translates to ~33% MAC overhead reduction, and hence requiring less airtime for their transmission.

The format of the WUR Short Wake-up frame is shown in Figure 7.27.

The Frame Control field is only 4 bits long and contains the Type subfield, which identifies the WUR Short Wake-up frame and the Protected subfield, which is set to 0 as these frames are not MIC protected.

This is an individually addressed frame sent to a WUR STA that supports its reception. The WUR ID field contains the identifier of the WUR STA, which is the intended recipient of the frame.

The FCS field contains the CRC, wherein the *calculation fields* include the Embedded BSSID field as well (i.e. the frame can be received successfully by a WUR STA that already knows the BSSID of the WUR BSS).

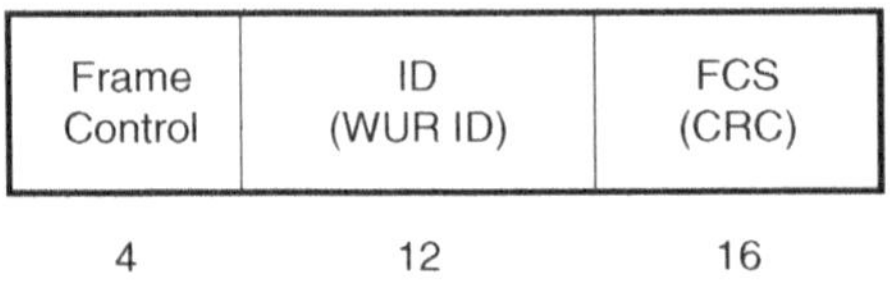

Frame Control	ID (WUR ID)	FCS (CRC)
4	12	16

Figure 7.27 WUR Short Wake-up frame format.

7.5   Conclusion

In this chapter, we have covered some details of various IEs, MR MAC frames, and WUR MAC frames that are essential to understand the functionalities discussed in previous chapters. If a reader is interested in learning more details about the functionalities, this chapter may also be used as a quick reference for the reader to explore the IEEE 802.11ba and 802.11 standards.

Index

a

Access point   2
Active mode   21, 105
Active scanning   16, 97
AP, *See* Access point
Association Request frame   18, 156
Association Response frame
   18, 156
Awake state   21, 105

b

Basic service set identifier   99
Beacon   16, 155
Beacon interval   16
BSSID, *See* Basic service set
   identifier
BU, *See* Bufferable unit
Bufferable unit   116

c

Calculation fields   161
Carrier sense multiple access with
   collision avoidance   19
Channel Model D   78
CRC, *See* Cyclic redundancy check

Critical update   106, 121
CSMA/CA, *See* Carrier sense
   multiple access with collision
   avoidance
Cyclic redundancy check   130

d

Data field   61–62
DC, *See* Duty cycle
DCF, *see* Distributed coordination
   function
Deep sleep mode   30
Delivery context   117
Delivery TIM   106
DFS, *See* Dynamic frequency
   selection
Distributed coordination function   19
Doze state   21, 105
DTIM, *See* Delivery TIM
Duty cycle   32, 112
Dynamic frequency selection   67

e

EDCA, *See* Enhanced distributed
   channel access

IEEE 802.11ba: Ultra-Low Power Wake-up Radio Standard, First Edition.
Steve Shellhammer, Alfred Asterjadhi, and Yanjun Sun.
© 2023 The Institute of Electrical and Electronics Engineers, Inc.
Published 2023 by John Wiley & Sons, Inc.

Element   131
Enhanced distributed channel access
   20, 132

f

FCS, *See* Frame check sequence
FDMA operation   66–67, 127
FL WUR Wake-up frame   116, 162
Frame check sequence   130

g

Group addressed   119

h

HDR, *See* High data rate
High data rate   61–62

i

IEEE, *See* Institute of Electrical and
   Electronic Engineers
IEEE 802.11   1, 9–24
IEEE 802.11ba   1
Individually addressed   116
Institute of Electrical and Electronic
   Engineers   1
Integrity check   129

k

Keep alive frame   126

l

LDR, *See* Low data rate
Least significant bits   99
Link budget   50–52, 92–94
LNA, *See* Low noise amplifier
Long training field   13
Low data rate   61–62
Low noise amplifier   28
LSB, *See* Least significant bits
LTF, *See* Long training field

m

MAC, *See* Medium access control
Main radio   2, 45
MC-OOK, *See* Multicarrier on-off
   keying
Medium access control   97
Message integrity check   130
MIC, *See* Message integrity check
Most significant bits   99
MR, *See* Main radio
MSB, *See* Most significant bits
Multicarrier on-off keying   53–54

n

Network discovery   16, 97
Noise figure   28
Non-coherent OOK   51
Non-transmitted BSSID   103

o

OFDM, *See* Orthogonal frequency
   division multiplexing
OFDMA, *See* Orthogonal frequency
   division multiple access
OOK symbol   103
Operating Frequencies   10–11
Orthogonal frequency division
   multiple access   23
Orthogonal frequency division
   multiplexing   11

p

Packet error rate   88–92
PAR, *See* Project authorization
   request
Partial TSF   102
Passive scanning   16, 97
PER, *See* Packet error rate
PHY Protocol Data Unit   12,
   55–67

PN, *See* Pseudorandom noise
Power amplifier 27–28
Power consumption 26
Power management 21, 105
Power save mode 21, 105
PPDU, *See* PHY Protocol Data Unit
Primary channel 102
Probe Request frame 17, 156
Probe Response frame 17, 156
Project authorization request 45
Protected WUR frame 129
Pseudorandom noise 130
PS mode, *See* Power save mode

r
Radiolocation radar systems 102
Regulations 47–50

s
Security 20, 129
Service set identifier 16, 99
Short training field 13
SIG, *See* Signal field
Signal field 13
Sleep mode 30–31
SSID, *See* Service set identifier
STA 4
STF, *See* Short training field
Subcarrier 11
Symbol randomization 62–65
Sync field 58–61
Sync field classification 83–84
Sync field detection 82–83
Sync field timing 85–87

t
Target beacon transmission
time 21
Target WUR beacon transmit
times 102

TBTT, *See* Target beacon transmission
time
TIM element, *See* Traffic
indication map
Timing synchronization function
21, 102
Traffic indication map 22
Transmitted BSSID 103
TSF, *See* Timing synchronization
function
TWBTT, *See* Target WUR beacon
transmit times

v
VL WUR Wake-up frame 116

w
Wake-up radio 1
Wake-up radio integrity group
temporal key 130
Wake-up radio temporal key 130
WIGTK, *See* Wake-up radio integrity
group temporal key
WTK, *See* Wake-up radio temporal
key
WUR, *See* Wake-up radio
WUR Beacon frame 102, 161
WUR Beacon period 102
WUR Capabilities element 142
WUR DC, *See* WUR duty cycle
WUR discovery 98
WUR discovery channel 98
WUR Discovery element 154
WUR Discovery frame 99, 164
WUR discovery period 99
WUR duty cycle 113
WUR frame 157
WUR mode 108
WUR Mode element 108, 145
WUR Operation element 104, 142

WUR PN Update element 110, 155
WUR Power management 106
WUR Power State 107
WUR primary channel 102

WUR Short Wake-up frame
 124, 166
WUR wake up 116
WUR Wake-up frame 161

Printed and bound by CPI Group (UK) Ltd, Croydon, CR0 4YY